J. P. LaSalle

Applied Mathematical Sciences 62

The Stability and Control of Discrete Processes

Springer-Verlag

Applied Mathematical Sciences

EDITORIAL STATEMENT

The mathematization of all sciences, the fading of traditional scientific boundaries, the impact of computer technology, the growing importance of mathematical-computer modelling and the necessity of scientific planning all create the need both in education and research for books that are introductory to and abreast of these developments.

The purpose of this series is to provide such books, suitable for the user of mathematics, the mathematician interested in applications, and the student scientist. In particular, this series will provide an outlet for material less formally presented and more anticipatory of needs than finished texts or monographs, yet of immediate interest because of the novelty of its treatment of an application or of mathematics being applied or lying close to applications.

The aim of the series is, through rapid publication in an attractive but inexpensive format, to make material of current interest widely accessible. This implies the absence of excessive generality and abstraction, and unrealistic idealization, but with quality of exposition as a goal.

Many of the books will originate out of and will stimulate the development of new undergraduate and graduate courses in the applications of mathematics. Some of the books will present introductions to new areas of research, new applications and act as signposts for new directions in the mathematical sciences. This series will often serve as an intermediate stage of the publication of material which, through exposure here, will be further developed and refined. These will appear in conventional format and in hard cover.

MANUSCRIPTS

The Editors welcome all inquiries regarding the submission of manuscripts for the series. Final preparation of all manuscripts will take place in the editorial offices of the series in the Division of Applied Mathematics, Brown University, Providence, Rhode Island.

SPRINGER-VERLAG NEW YORK INC., 175 Fifth Avenue, New York, N. Y. 10010

Applied Mathematical Sciences | Volume 62

Applied Mathematical Sciences

1. John: **Partial Differential Equations,** 4th ed.
2. Sirovich: **Techniques of Asymptotic Analysis.**
3. Hale: **Theory of Functional Differential Equations,** 2nd ed.
4. Percus: **Combinatorial Methods.**
5. von Mises/Friedrichs: **Fluid Dynamics.**
6. Freiberger/Grenander: **A Short Course in Computational Probability and Statistics.**
7. Pipkin: **Lectures on Viscoelasticity Theory.**
9. Friedrichs: **Spectral Theory of Operators in Hilbert Space.**
11. Wolovich: **Linear Multivariable Systems.**
12. Berkovitz: **Optimal Control Theory.**
13. Bluman/Cole: **Similarity Methods for Differential Equations.**
14. Yoshizawa: **Stability Theory and the Existence of Periodic Solutions and Almost Periodic Solutions.**
15. Braun: **Differential Equations and Their Applications,** 3rd ed.
16. Lefschetz: **Applications of Algebraic Topology.**
17. Collatz/Wetterling: **Optimization Problems.**
18. Grenander: **Pattern Synthesis: Lectures in Pattern Theory, Vol I.**
20. Driver: **Ordinary and Delay Differential Equations.**
21. Courant/Friedrichs: **Supersonic Flow and Shock Waves.**
22. Rouche/Habets/Laloy: **Stability Theory by Liapunov's Direct Method.**
23. Lamperti: **Stochastic Processes: A Survey of the Mathematical Theory.**
24. Grenander: **Pattern Analysis: Lectures in Pattern Theory, Vol. II.**
25. Davies: **Integral Transforms and Their Applications,** 2nd ed.
26. Kushner/Clark: **Stochastic Approximation Methods for Constrained and Unconstrained Systems.**
27. de Boor: **A Practical Guide to Splines.**
28. Keilson: **Markov Chain Models—Rarity and Exponentiality.**
29. de Veubeke: **A Course in Elasticity.**
30. Sniatycki: **Geometric Quantization and Quantum Mechanics.**
31. Reid: **Sturmian Theory for Ordinary Differential Equations.**
32. Meis/Markowitz: **Numerical Solution of Partial Differential Equations.**
33. Grenander: **Regular Structures: Lectures in Pattern Theory, Vol. III.**
34. Kevorkian/Cole: **Perturbation Methods in Applied Mathematics.**
35. Carr: **Applications of Centre Manifold Theory.**
36. Bengtsson/Ghil/Källén: **Dynamic Meterology: Data Assimilation Methods.**
37. Saperstone: **Semidynamical Systems in Infinite Dimensional Spaces.**
38. Lichtenberg/Lieberman: **Regular and Stochastic Motion.**

(continued on inside back cover)

J.P. LaSalle

The Stability and Control of Discrete Processes

With 15 Illustrations

Springer-Verlag
New York Berlin Heidelberg
London Paris Tokyo

AMS Subject Classifications: 34A30, 34D99, 34H05, 49E99, 58F10, 93C05, 93C55

Library of Congress Cataloging in Publication Data
LaSalle, Joseph P.
The stability and control of discrete processes.
(Applied mathematical sciences; v. 62)
Bibliography: p.
Includes index.
1. Control theory. 2. Stability. 3. Differential equations. I. Title. II. Series: Applied mathematical sciences (Springer-Verlag New York Inc.); 62.
QA1.A647 vol. 62 510 s 86-17757
[QA402.3] [629.8′312]

Printed and bound by Quinn–Woodbine, Inc., Woodbine, New Jersey.
Printed in the United States of America.

9 8 7 6 5 4 3 2 1

ISBN 0-387-96411-8 Springer-Verlag New York Berlin Heidelberg
ISBN 3-540-96411-8 Springer-Verlag Berlin Heidelberg New York

FOREWORD

Professor J. P. LaSalle died on July 7, 1983 at the age of 67. The present book is being published posthumously with the careful assistance of Kenneth Meyer, one of the students of Professor LaSalle. It is appropriate that the last publication of Professor LaSalle should be on a subject which contains many interesting ideas, is very useful in applications and can be understood at an undergraduate level. In addition to making many significant contributions at the research level to differential equations and control theory, he was an excellent teacher and had the ability to make sophisticated concepts appear to be very elementary. Two examples of this are his books with N. Hasser and J. Sullivan on analysis published by Ginn and Co., 1949 and 1964, and the book with S. Lefschetz on stability by Liapunov's second method published by Academic Press, 1961. Thus, it is very fitting that the present volume could be completed.

Jack K. Hale
Kenneth R. Meyer

TABLE OF CONTENTS

1. Introduction

This book will discuss the stability and controllability of a discrete dynamical system. It is assumed that at any particular time the system can be completely described by a finite dimensional vector $x \in \mathbb{R}^m$ -- the state vector. Here $\mathbb{R}^m$ is the real m-dimensional Euclidean space and

$$x = \begin{pmatrix} x_1 \\ \cdot \\ \cdot \\ \cdot \\ x_m \end{pmatrix} \in \mathbb{R}^m, \quad ||x|| = (x_1^2 + \dots + x_m^2)^{1/2},$$

(the Euclidean length of the vector x). The components $x_1, \dots, x_m$ might be the temperature, density, pressure etc. of some physical system.

The system is observed only at discrete times; e.g., every hour or every second. Let J denote the set of all integers and J_+ the set of all nonnegative integers, thus the future of the system would be a sequence $\{x^j\}$ where $j \in J_+$ and $x^j \in \mathbb{R}^m$. For a sequence of vectors x^j, $x^j \to y$ as $j \to \infty$ means $||x^j - y|| \to 0$ as $j \to \infty$. We will also use x to denote a function on J_+ to $\mathbb{R}^m$ $(x: J_+ \to \mathbb{R}^m)$. Thus we allow the usual and convenient ambiguity where x may denote either a function or a vector (point).

The <u>first difference</u> $\dot{x}$ of a function x is defined by $\dot{x}(n) = x(n+1) - x(n)$. Corresponding to the fundamental theorem of calculus, we have $\sum_{k=j}^{n} \dot{x}(k) = x(n+1) - x(j)$ and, if $y(n+1) = \sum_{k=j}^{n} x(k)$, then $\dot{y}(n) = x(n)$. These are two of the basic formulas of the finite calculus (the calculus of finite differences) which has been studied as long as the continuous (infinitesimal) calculus and dates back to Brook Taylor (1717) and Jacob Stirling (1730). We use x' to denote the function defined by $x'(n) = x(n+1)$, so $x' = \dot{x} + x$.

The fundamental assumption that we shall make is that the system we wish to describe obeys a difference equation. That is, there is a function $T: \mathbf{R}^m \to \mathbf{R}^m$ such that

$$x(n+1) = T(x(n)) \quad \text{for} \quad n \in J_+. \tag{1.1}$$

Thus the state at time $n+1$ is completely determined by the state at time n. Sometimes we shall allow the function T to depend on n. A function $T: \mathbf{R}^m \to \mathbf{R}^m$ is continuous (on $\mathbf{R}^m$) if $x^j \to y$ implies $T(x^j) \to T(y)$. We do not need to do so now but later will always assume that T is continuous.

1.1. Exercise. Show that if $x(n+1) = 3\,x(n)$ for $n \in J_+$ where x is a scalar and $x(0) = 7$ then $x(n) = 7 \cdot 3^n$.

If x is a vector, $T(x)$ is a vector and is the value of the function at x. If x is a function, the symbol Tx will denote the composition of functions -- $(Tx)(n) = T(x(n))$. Thus the difference equation (1.1) can be written

$$x' = Tx. \tag{1.2}$$

The solution to the initial value problem

$$x' = Tx, \quad x(0) = x^0, \tag{1.3}$$

is $x(n) = T^n(x^0)$, where T^n is the nth-iterate of T. That is $T^0 = I$, the identity function $(Ix = x)$, and $T^n = T \cdot T^{n-1}$. The solution is defined on J_+. There are no difficult questions about the existence and uniqueness of solutions for difference equations. Also, it is clear that the solutions are continuous with respect to the initial condition (state) x^0 if T is continuous. Unlike ordinary differential equations, the existence and uniqueness is only in the forward direction of time $(n \geq 0)$. The equation (1.3) is simply an algorithm defining a function x on J_+.

Let $g: \mathbf{R}^m \to \mathbf{R}^1$. Then

$$u(n+m) = g(u(n), u(n+1),\ldots,u(n+m-1)) \tag{1.4}$$

is an mth-order difference equation. The value of u at time $n+1$ depends on its value at times $n, n-1,\ldots,n-m+1$ - i.e., upon this portion of its past history. Note that this equation can be written in the form (1.2) by defining $x_1(n) = u(n)$, $x_2 = x_1'(n) = u(n+1),\ldots,x_m(n) = x_{m-1}'(n) = u(n+m-1)$ and so (1.4) is equivalent to the following system of m first-order difference equations (the vector $x(n)$ is the state of (1.4) at time $n+m-1$):

$$\begin{aligned} x_1' &= x_2 \\ x_2' &= x_3 \\ &\vdots \\ x_{m-1}' &= x_m \\ x_m' &= g(x_1,\ldots,x_m) = g(x) \end{aligned} \tag{1.5}$$

or

$$x' = Tx,$$

where

$$x = \begin{pmatrix} x_1 \\ \vdots \\ x_{m-1} \\ x_m \end{pmatrix}, \qquad T(x) = \begin{pmatrix} x_2 \\ \vdots \\ x_m \\ g(x) \end{pmatrix}.$$

For instance, the 3rd order linear equation

$$u''' + a_2u'' + a_1u' + a_0u = 0$$

is equivalent to

$$x' = Ax$$

where

$$x_1 = u,\ x_2 = u',\ x_3 = u'',\ x = \begin{pmatrix} x_1 \\ x_2 \\ x_3 \end{pmatrix}$$

and

$$A = \begin{pmatrix} 0 & 1 & 0 \\ 0 & 0 & 1 \\ -a_0 & -a_1 & -a_2 \end{pmatrix}.$$

a 3×3 matrix. The initial conditions $u(0) = u_0$, $u'(0) = u_0'$ and $u''(0) = u_0''$ correspond to

$$x(0) = \begin{pmatrix} u_0 \\ u_0' \\ u_0'' \end{pmatrix} = x^0.$$

The solution of this initial value problem is

$$x(n) = A^n x^0,$$

where A^n is the product of the matrix A with itself n times. We are often interested in obtaining information about the asymptotic behavior of solutions; i.e., what happens to solutions for large n? For instance, in the above example, when does $A^n \to 0$ as $n \to \infty$ ($A^n = (a_{ij}^{(n)})$ and $A^n \to 0$ means $a_{ij}^{(n)} \to 0$ for all i,j or, equivalently $A^n x^0 \to 0$ for each x^0).

1.2. Exercise. The Fibonacci numbers $\{f^n\}$, $n \geq 0$ are defined by $f^n = f^{n-1} + f^{n-2}$ for $n \geq 2$, $f^0 = 0$, $f^1 = 1$. Compute the first 6 Fibonacci numbers using the definition given above. Write the equivalent system of two first order difference equations.

A state $x^0 \in \mathbf{R}^m$ is fixed point or equilibrium state for (1.2) if $T(x^0) = x^0$. Thus the solution of (1.2) which starts at x^0 stays at x^0 for all $n \in J_+$. A solution $T^n x^0$ is said to be periodic (or cyclic) if for some $k > 0$, $T^k x^0 = x^0$. The least such k is called the period of the solution or the order of the cycle. If $k = 1$ then x^0 is a fixed point.

Note here, unlike an ordinary differential equation, a solution can reach an equilibrium in finite time.

1.3. Exercise. Consider $x' = 4x(1-x)$. Show that 0 and 3/4 are the only fixed points. Also note that if $x^0 = 1$ then $x^n = 0$ for $n \geq 1$.

1.4. Exercise. Consider $x_1' = x_2$, $x_2' = -x_1$. Show that every solution is periodic with period 4.

A subset $A \in \mathbf{R}^m$ is said to be positively invariant if $T(A) \subset A$, i.e., if $x \in A$ then $T(x) \in A$. Clearly an equilibrium solution is invariant. The set $\{T^n(x_0)\colon n \in J_+\}$ is called the positive orbit of x_0 and is positively invariant.

1.5. Exercise. Show that the positive orbit of a periodic orbit is a finite invariant set.

1.6. Exercise. Show that $|x_1| < 5$ is a positively invariant set for $x_1' = (1/2)x_1$.

1.7. Exercise. Consider the system $x' = Ax$ where $A = \begin{pmatrix} \cos\alpha & -\sin\alpha \\ \sin\alpha & \cos\alpha \end{pmatrix}$ and α is a constant. Show that in polar coordinates this difference equation becomes $r' = r$, $\Theta' =$

$\Theta + \alpha$. Thus circles are invariant.

Consider (1.2) again and a particular solution $T^n(x_0)$. A point $y \in \mathbf{R}^m$ is a <u>positive limit point</u> of $T^n(x_0)$ if there is a subsequence n_i with $T^{n_i}(x_0) \to y$. The <u>positive limit set</u> (or ω-<u>limit set</u>) $\omega(x_0)$ of $T^n(x_0)$ is the set of all positive limit points of $T^n(x_0)$. A sequence $x^n \in \mathbf{R}^m$ is said to approach a set $A \subset \mathbf{R}^m$ if $\rho(x^n,A) = \inf\{||x^n-y||: y \in A\}$.

1.8. <u>Exercise</u>. Consider the difference equation in polar coordinates given by $r' = r + rg(r)$, $\theta' = \Theta + \alpha$ where g is continuous for $r \geq 0$ and α is a constant. Show that in rectangular coordinates the equations are $x' = Ax(1 + g(||x||))$ where A is the 2×2 matrix in Exercise 1.7.

1.9. <u>Exercise</u>. Let $\alpha = \pi/2$, $x^0 = (1/2,0)^T$, $g(r) = 1-r$ for the equation in Exercise 1.8. Show that $\omega(x^0) = \{(1,0), (0,1), (-1,0), (0,-1)\}$.

1.10. <u>Proposition</u>. If $T^n(x^0)$ is bounded for all $n \in J_+$ then $\omega(x^0)$ is a non-empty, compact positively invariant set. Moreover, $T^n(x^0) \to \omega(x^0)$ as $n \to \alpha$.

<u>Proof</u>: Let $\omega = \omega(x^0)$ and $x^n = T^n(x^0)$. It is easy to see that the complement of ω is open and so ω is closed. Since x^n is bounded, $||x^n|| \leq M$ where M is fixed, it is clear that ω is bounded, $||y|| \leq M$ for $y \in \omega$. By the Heine-Borel theorem, ω is compact. Since x^n is bounded the Bolzano-Weierstrass theorem asserts the existence of at least one limit point and so ω is non-empty.

Let $y \in \omega$ so $x^{n_i} \to y$. Since T is continuous $x^{n_i+1} = T(x^{n_i}) \to T(y)$ or $T(y) \in \omega$. This ω is positively invariant. $\rho(x^n,\omega)$ is bounded since both x^n and ω are. Assume $\rho(x^n,\omega)$ does not converge to zero. Then there is a subsequence n_i such that $x^{n_i} \to y$ and $\rho(x^{n_i},\omega) \to a > 0$. But then $y \in \omega$, $\rho(x^{n_i},\omega) \leq (x^{n_i},y) \to 0$ so $\rho(x^{n_i},\omega) \to 0$ which is a contradiction. □

2. Liapunov's Direct Method

In this section, we introduce the concept of Liapunov function in order to discuss stability questions for difference equations. Consider the difference equation

$$x' = Tx, \quad x(0) = x^0 \tag{2.1}$$

where as before $T: \mathbf{R}^m \to \mathbf{R}^m$ is continuous. Throughout this section, we assume $\bar{x}$ is an equilibrium solution for (2.1) so $T(\bar{x}) = \bar{x}$ and $x(n) = \bar{x}$ is a solution.

If $\delta > 0$ then the δ-ball about a point $x^0 \in \mathbf{R}^m$ is defined as $B_\delta(x^0) = \{x \in \mathbf{R}^m: ||x-x^0|| < \delta\}$. We say that $\bar{x}$ is a stable solution if for every $\varepsilon > 0$ there is a $\delta > 0$ such that $T^n(B(\bar{x})) \subset B_\varepsilon(\bar{x})$ (i.e., $||T^n(x) - \bar{x}|| < \varepsilon$ whenever $||x-\bar{x}|| < \delta$) for all $n \geq 0$.) This says that you can stay as close to the equilibrium for all future times provided you start sufficiently close. If the equilibrium solution $\bar{x}$ is not stable then it is called unstable.

2.1. Exercise. a) Show that the origin is stable for $x_1' = x_2$, $x_2' = -x_1$ and b) the origin is unstable for $x_1' = 2x_1$.

Let $V: \mathbf{R}^m \to \mathbf{R}$. Relative to (2.1) (or to T) define

$$\dot{V}(x) = V(T(x)) - V(x).$$

If $x(n)$ is a solution of (2.1)

$$\dot{V}(x(n)) = V(x(n+1)) - V(x(n));$$

$\dot{V}(x) \leq 0$ means that V is non-increasing along solutions. Henceforth, we shall assume $V(\bar{x}) = 0$.

Let G be any set in $\mathbf{R}^n$. We say that V is a Liapunov function for (2.1) on G if (i) V is continuous on $\mathbf{R}^n$ and (ii) $\dot{V}(x) \leq 0$ for all $x \in G$. A function V is said to be positive definite with respect to $\bar{x}$ if (i) $V(\bar{x}) = 0$ and (ii) there is an $\eta > 0$ such that $V(x) > 0$ wherever $x \in B_\eta(\bar{x})$, $x \neq \bar{x}$.

2.2. Proposition (Liapunov's Stability Theorem). If V is a Liapunov function for T on some neighborhood of $\bar{x}$ and V is positive definite with respect to $\bar{x}$ then $\bar{x}$ is a stable equilibrium.

This is the first of several theorems due to the turn-of-the-century, Russian mathematician A. M. Liapunov. These theorems constitute his "direct method" since they establish stability, instability, etc. without specific knowledge of the solution.

Proof: We may take η so small that $V(x) > 0$ and $\dot{V}(x) \leq 0$ for $x \in B_\eta(x)$. Let $\varepsilon > 0$ be given; there is no loss in generality in taking $0 < \varepsilon < \eta$. Let $m = \min\{V(x): ||x-\bar{x}|| = \varepsilon\}$. m is positive since we are taking the minimum of a positive continuous function over a compact set. Let $G = \{x | V(x) < m/2\}$ and G_0 the connected component of G which contains $\bar{x}$. Both G and G_0 are open. If $x^0 \in G_0$ then $\dot{V}(x^0) \leq 0$ so $V(T(x^0)) \leq V(x^0) < m/2$ so $x^0 \in G$. Since x^0 and $\bar{x}$ are in the same component of G so are $T(\bar{x}) = \bar{x}$ and $T(x^0)$. Thus G_0 is an open positively invariant set containing $\bar{x}$ and contained in $B_\varepsilon(\bar{x})$. Since V is continuous there is a $\delta > 0$ such that $B_\delta(\bar{x}) \subset G_0$. So if $x^0 \in B_\delta$, then $x^0 \in G$

and $T(x^0) \subset G_0 \subset B_\varepsilon(\bar{x})$. □

We say the solution through x^0 is *bounded* provided there is a constant M such that $||T^n(x^0)|| \leq M$ for all $n \geq 0$. Using the same type of argument you can prove:

2.3. *Proposition*. *If V is a Liapunov function for (2.1) on the set where $||x||>N$ (a constant) and $V(x) \to \infty$ as $x \to \infty$ then all solutions of (2.1) are bounded.*

2.4. *Exercise*. Show that the origin is stable and all solutions are bounded for the difference equation $x_1' = x_2$, $x_2' = -x_1$ by considering the function $V = x_1^2 + x_2^2$.

The equilibrium solution $\bar{x}$ is *asymptotically stable* if (i) it is stable and (ii) there exists an $\eta > 0$ such that if $x^0 \in B_\eta(\bar{x})$ then $\lim_{n\to\infty} T^n(x^0) = \bar{x}$.

2.5. *Exercise*. (i) Show that the origin is asymptotically stable for $x_1' = (1/2)x_1$. (ii) Show that the origin is stable but not asymptotically stable for $x_1' = x_2$, $x_2' = -x_1$.

We have a result which establishes stability, so what we need is a separate result which proves that the solutions tend to the origin. This type of result can be obtained from the discussion of the ω-limit sets of an orbit found in the previous section.

2.6. *Proposition*. (Invariance Principle) *If* (i) V *is a Liapunov function for* (2.1) *in* G *and* (ii) *the solution* $T^n(x^0)$ *of* (2.1) *is in* G *and bounded, then there is a number* c *such that* $T^n(x^0) \to M \cap V^{-1}(c)$ *where* M *is the largest positively invariant set contained in the set* $E = \{x \in \mathbb{R}^m: \dot{V} = 0\} \cap \bar{G}$.

Remark. The above theorem is known as LaSalle's theorem and the set E is called the LaSalle set (KRM).

Proof: Since $x^n = T^n(x^0)$ is bounded and in G we have $\omega = \omega(x^0) \neq \phi$, $\omega \subset G^-$ and x_n tends to ω. Now $V(x^n)$ is non-increasing and bounded below, so $V(x^n) \to c$. If $y \in \omega$ then there is a subsequence n_i such that $x^{n_i} \to y$ so $V(x^{n_i}) \to V(y)$ or $V(y) = c$. Thus $V(\omega) = c$ or $\omega \subset V^{-1}(c)$. Also since $V(\omega) = c$ and ω is positively invariant $\dot{V}(\omega) = 0$ So $x^n \to \omega \subset \{x \in \mathbb{R}^m : \dot{V} = 0\} \cap G^- \cap V^{-1}(c)$. Since ω is positively invariant we have $\omega \subset M$. □

The difficulty in applications is to find a "good" Liapunov function - one that makes M as small as possible. For instance a constant function is a Liapunov function but gives no information.

Let us look at a simple example which illustrates how the result is applied. Consider the 2-dimensional system

$$x' = \frac{ay}{1+x^2}, \qquad y' = \frac{bx}{1+y^2}$$

(here $(x_1,x_2) = (x,y)$ and a, b are constants.) Take $V = x^2 + y^2$ so

$$\dot{V} = (x')^2 + (y')^2 - (x^2 + y^2)$$

$$\left(\frac{b^2}{(1+y^2)^2} - 1\right)x^2 + \left(\frac{a^2}{(1+x^2)^2} - 1\right)y^2$$

$$\leq (b^2-1)x^2 + (a^2-1)y^2$$

When $a^2 \leq 1$ and $b^2 \leq 1$ then V is a Liapunov function on all of R^2. Since V is positive definite with respect to the origin the origin is stable by Proposition 2.2 and since $V \to \infty$ as $x^2 + y^2 \to \infty$ all solutions are bounded by Proposition 2.3.

Case 1. $a^2 < 1$, $b^2 < 1$. In this case $M = E = \{(0,0)\}$ and so all solutions tend to the origin. (When all solutions are bounded, the origin is stable and all solutions tend to the origin then the origin is said to be globally asymptotically stable.)

Case 2. $a^2 \leq 1$, $b^2 \leq 1$ but $a^2 + b^2 \neq 2$. In this case we may assume $a^2 < 1$ and $b^2 = 1$. In this case $\dot{V} \leq (a^2-1)y^2$ and $E = \{(x,0)\}$ the x-axis. Now $T(x,0) = (0,bx) = (0,x)$ so the only invariant subset of E is the origin. Thus we still have global asymptotic stability.

Case 3. $a^2 = b^2 = 1$. V is still a Liapunov function and $E = M$ is the union of the x and y axes.

By Proposition 2.6 all solutions tend to $\{(c,0),$ $(-c,0),$ $(0,c),$ $(0,-c)\}$, the intersection of E and the circle $V = c^2$. In the case when $ab = +1$ this set consists of two periodic solutions of period 2 and when $ab = -1$ this set consists of one periodic solution of period 4. A more detailed discussion can be found in LaSalle [1].

2.7. Corollary. If V and $-\dot{V}$ are positive definite with respect to $\bar{x}$, then $\bar{x}$ is asymptotically stable. (This is the classical Liapunov theorem on asymptotic stability.)

Proof: Since $-\dot{V} > 0$ on a neighborhood of $\bar{x}$ the point $\bar{x}$ is stable by Proposition 2.2. From the proof of Proposition 2.2 there is an arbitrary small neighborhood G_0 of $\bar{x}$ which is positively invarant. We can make G_0 so small that $V(x) > 0$ and $\dot{V}(x) < 0$ for $x \in G\{\bar{x}\}$. Given any $x^0 \in G_0$ we have by the invariance principle that $T^n(x^0)$ tends to the largest invariant set in $G_0 \cap \{\dot{V}(x) = 0\} = \bar{x}$ since $-\dot{V}$ is positive definite. Thus, $\bar{x}$ is asymptotically stable.

2.8. Proposition. *Let $\dot{V}$ be positive definite with respect to $\bar{x}$ and let V take positive values arbitrarily close to $\bar{x}$, then $\bar{x}$ is unstable.*

Proof: Assume to the contrary that $\bar{x}$ is stable. Let $\varepsilon > 0$ be so small that $\dot{V}(x) > 0$ for $x \in B_\varepsilon(\bar{x}) \setminus \{\bar{x}\}$ and $\delta > 0$ so that if $x^0 \in B_\delta(\bar{x})$ then $x^n = T^n(x_0) \in B_\varepsilon(\bar{x})$ for all n. By the hypothesis there is a point $x^0 \in B_\delta(\bar{x})$ such that $V(x^0) > 0$. Since x^n is bounded and remains in $B_\varepsilon(x)$, x^n tends to $\bar{x} = \{x \mid \dot{V}(x) = 0\} \cap B_\varepsilon(\bar{x})$. Since $x^n \to \bar{x}$ we have $V(x^n) \to V(0) = 0$. But $\dot{V}(x^n) > 0$ so $V(x^n) \geq 0$ so $V(x^n) \geq V(x^{n-1}) \geq \ldots \geq V(x^0) > 0$. This contradiction proves the theorem. □

In order to prove an important theorem on instability by the first approximation (Proposition 7.1) we shall need a slight different result.

2.9. Proposition. Assume V takes positive values arbitrarily close to $\bar{x}$ and $\dot{V} = \beta V + W$ where $W(x) \geq 0$ on some neighborhood of $\bar{x}$ and $\beta > 1$, then the origin is unstable.

Proof: Again assume that x is stable and ε and δ as in the above proof. Choose $x^0 \in B_\delta(\bar{x})$ where $V(x^0) > 0$. Since $x^n = T^n(x_0) \in B_\varepsilon(\bar{x})$ where $W(x) \geq 0$ we have $\dot{V}(x^n) > \beta V(x^n)$. By induction $v(x^n) > \beta^n V(x^0)$ and so $v(x^n) \to \infty$ as $n \to \infty$. But this contradicts the fact that x^n is bounded and V is continuous. □

3. Linear systems $x' = Ax$.

3.1. Some preliminary definitions and notations.

$$B = (b_{ij}) = \begin{pmatrix} b_{11} & b_{12} & \cdots & b_{1s} \\ b_{21} & b_{22} & \cdots & b_{2s} \\ \vdots & & & \\ b_{r1} & b_{r2} & \cdots & b_{rs} \end{pmatrix} = (b^1 b^2 \ldots b^s)$$

is an $r \times s$ matrix (real or complex), where

$b^j = \begin{pmatrix} b_{1j} \\ b_{2j} \\ \vdots \\ b_{rj} \end{pmatrix}$ is the $j^{\underline{th}}$-column vector in B. Thus, for

$c = (c_i) = \begin{pmatrix} c_1 \\ \vdots \\ c_s \end{pmatrix}$ any s-vector

$$Bc = c_1 b^1 + c_2 b^2 + \cdots + c_s b^s.$$

$\|B\|$, the <u>norm</u> of B, is defined by

$$\|B\| = \max\{\|Bc\|;\ \|c\| = 1,\ c \in \mathbf{C}^s\}.$$

Note (Exercise 3.3), if B is real, it makes no difference in the definition of $\|B\|$ whether c ranges over $\mathbf{R}^s$ or $\mathbf{C}^s$.

$$B^T = (b_{ji}) = \begin{pmatrix} b_{11} & b_{21} & \cdots & b_{r1} \\ b_{12} & b_{22} & \cdots & b_{r2} \\ \vdots & \vdots & & \vdots \\ b_{1s} & b_{2s} & & b_{rs} \end{pmatrix} = \begin{pmatrix} b^1 \\ b^2 \\ \vdots \\ b^s \end{pmatrix}$$

is the transpose of B.

For $X(n)$, $n \geq 0$, a sequence of $r \times s$ matrices, $X(n) \to C$ as $n \to \infty$ if $\|X(n)-C\| \to 0$ as $n \to \infty$; $X(n)$ is said to be <u>bounded</u> if $\|X(n)\|$ is bounded for all $n \geq 0$.

3.2. <u>Exercise</u>. Show that: $||B|| = \sup\{|Bc|;\ |c| \le 1,\ c \in \mathbf{C}^s\}$.

3.3. <u>Exercise</u>. If B is real, show that

$$||B|| = \sup\{||Bc||;\ ||c|| = 1,\ c \in \mathbf{R}^s\}.$$

3.4. <u>Exercise</u>. Show that:

(i) $||B|| \ge 0$ and $||B|| = 0$ if and only if $B = 0$.

(ii) $||B+C|| \le ||B|| + ||C||$

(iii) $||BC|| \le ||B||\,||C||$,

whenever the operators are defined.

3.5. <u>Exercise</u>. Show that

a. $\max_j ||b^j|| \le ||B|| \le \sqrt{s}\, \max_j ||b^j||$.

b. For B real, $||B||^2$ is the largest eigenvalue of $B^T B$.

3.6. <u>Exercise</u>. If $X(n) \to C$ and $Y(n) \to D$ as $n \to \infty$, show that:

(i) $X(n) + Y(n) \to C + D$ as $n \to \infty$.

(ii) $X(n)Y(n) \to CD$ as $n \to \infty$,

whenever the operators are defined.

3.7. <u>Exercise</u>. Show that:

a. $X(n) \to C$ as $n \to \infty$ if and only if each $x_{ij}(n) \to c_{ij}$ as $n \to \infty$.

b. $X(n)$ is bounded if and only if each $x_{ij}(n)$ is bounded.

$A = (a_{ij})$ is a real $m \times m$ matrix.

$$\phi(\lambda) = \det(\lambda I - A) = \prod_{j=1}^{m} (\lambda - \lambda_j) = \lambda^m + \alpha_{m-1}\lambda^{m-1} + \cdots + \alpha_0$$

is the <u>characteristic polynomial of</u> A.

The λ_i are the <u>eigenvalues</u> of A and $\sigma(A) = \{\lambda_1,\ldots,\lambda_m\}$ is the <u>spectrum</u> of A.

$r(A) = \max_i |\lambda_i|$ is the <u>spectral radius</u> of A.

The <u>algebraic multiplicity</u> of λ_i is the multiplicity of λ_i as a root of $\phi(\lambda)$. Its <u>geometric multiplicity</u> is the dimension of the eigenspace $\{x;\ Ax = \lambda_i x\}$; i.e., the number of linearly independent eigenvectors associated with λ_i.

$A^0 = I$, $A^n = AA^{n-1}$, $n \geq 1$.

3.8. <u>Exercise</u>. Show that:

a. The geometric multiplicity m_i of λ_i is $m_i = m - \operatorname{rank}(A-\lambda_i I)$.

b. $r(A) \leq \|A\|$.

The general linear homogeneous system of difference equations of dimension m is

$$x_1(n+1) = a_{11}x_1(n) + a_{12}x_2(n) + \cdots + a_{1m}x_m(n)$$

$$x_2(n+1) = a_{21}x_1(n) + a_{22}x_2(n) + \cdots + a_{2m}x_m(n)$$

$$\vdots$$

$$x_m(n+1) = a_{m1}x_1(n) + a_{m2}x_2(n) + \cdots + a_{mm}x_m(n)$$

or

$$x' = Ax. \tag{3.1}$$

Since this is an autonomous system, we confine ourselves to solutions $x\colon J_0 \times R^m \to R^m$; i.e., to solutions of (3.1) that start at $n = 0$. The solution $\pi(n,x^0)$ of (3.1) satisfying $x(0) = x^0$ is

$$\pi(n,x^0) = A^n x^0. \tag{3.2}$$

The matrix $X(n) = A^n$ is the matrix solution of the matrix difference equation $X' = AX$ satisfying $X(0) = I$ and is called the <u>principal matrix solution</u>. The columns $x^j(n)$ of $X(n)$ are called the <u>principal solutions</u> of (3.1); x^j is the solution of (3.1) satisfying

$$x^j(0) = \begin{bmatrix} 0 \\ \vdots \\ 0 \\ 1 \\ 0 \\ \vdots \\ 0 \end{bmatrix} \longleftarrow j^{\text{th}} \text{ row.}$$

Thus

$$\pi(n,x^0) = x_1^0 x^1(n) + \cdots + x_m^0 x^m(n).$$

The space $\mathscr{S}$ of all solutions $x: J_0 \times \mathbf{R}^m \to \mathbf{R}^m$ of (3.1) is a linear space over $\mathbf{R}$, since the linear combination of solutions is also a solution. The principal solutions $x^1,\ldots,x^m$ span $\mathscr{S}$ and are clearly linearly independent, and hence

3.9. <u>Proposition</u>. The space $\mathscr{S}$ of all solutions of (3.1) starting at $n = 0$ is a finite dimensional linear space of dimension m.

Although we are interested only in real solutions of (3.1), many things are simplified by considering also the space $\hat{\mathscr{S}}$ of all complex solutions $x: J_0 \times \mathbf{C}^m \to \mathbf{C}^m$. Any such solution is given by $x(n) = A^n x(0)$. Hence the principal solutions are also a basis of $\hat{\mathscr{S}}$, and $\hat{\mathscr{S}}$ is also an m-dimensional linear space (over the complex field $\mathbf{C}$). From now on, if we do not say that a solution is real (or the context does not imply that it is real), we shall mean either a real or a complex solution. The matrix A is always assumed to be real but unless so specified other matrices may be real or complex. Part (b) of the exercise below shows that if we have a basis of complex solutions, we can always obtain from it a basis of real solutions.

3.10. Exercise. Show that:

a. Every basis of $\mathcal{S}$ is a basis of $\hat{\mathcal{S}}$.

b. If ξ^j, $j = 1,\ldots,m$, is a basis of $\hat{\mathcal{S}}$, then the real and imaginary parts of the ξ^j span $\hat{\mathcal{S}}$ and m of these real functions span $\hat{\mathcal{S}}$ (and are therefore a basis of $\mathcal{S}$).

From the above we can now make some simple observations concerning the stability of the origin for (3.1). The boundedness of all solutions of (3.1) is equivalent to the boundedness of the principal solutions of (3.1) (or of any basis of solutions), which is equivalent to the stability of the origin and the boundedness of A^n. Also, it is easy to see that asymptotic stability of the origin is always global, and that (i) $A^n \to 0$ as $n \to \infty$, (ii) all solutions approach the origin as $n \to \infty$, and (iii) the origin is asymptotically stable are equivalent. We shall say, if $A^n \to 0$ as $n \to \infty$, that A is stable. This is equivalent to asymptotic stability of the origin and is easier to say.

Let v^i be an eigenvector of A associated with an eigenvalue λ_i. Then $\lambda_i^n v^i$ is a solution of (3.1), and if $r(A) \geq 1$ there is always a solution that does not approach the origin as $n \to \infty$. If $r(A) > 1$, there is always an unbounded solution. Putting all this together, we have

3.11. Proposition. (a) The origin for (3.1) is stable if and only if A^n is bounded, and a necessary condition for this is that $r(A) \le 1$.

(b) The origin for (3.1) is asymptotically stable if and only if A is stable, and a necessary condition is $r(A) < 1$.

We will show in Section 5 that $r(A) < 1$ is also a sufficient condition for A to be stable. However, note that

$$x = \begin{pmatrix} n \\ n+1 \end{pmatrix}$$

is a solution of

$$\begin{aligned} x_1' &= x_2 \\ x_2' &= -x_1 + 2x_2, \end{aligned} \tag{3.3}$$

which shows that $r(A) \le 1$ is not a sufficient condition for the origin to be stable. (The characteristic equation for A is $(\lambda-1)^2$.)

3.12. Exercise. Determine the principal solutions of (3.3) and give a formula for A^n where

$$A = \begin{pmatrix} 0 & 1 \\ -1 & 2 \end{pmatrix}.$$

3.13. Exercise. Let

$$A = \begin{pmatrix} 1 & 1 & 1 & 1 & 1 \\ 1 & 1 & 1 & 1 & 1 \\ 1 & 1 & 1 & 1 & 1 \\ 1 & 1 & 1 & 1 & 1 \\ 1 & 1 & 1 & 1 & 1 \end{pmatrix}.$$

a. Show that $A(A-5I) = 0$.

b. Find the solution of $x' = Ax$ satisfying

$$x(0) = \begin{pmatrix} 1 \\ 1 \\ 1 \\ 1 \\ 1 \end{pmatrix}.$$

4. An algorithm for computing A^n.

The space of all $m \times m$ matrices is an m^2-dimensional linear space. In this section A is any $m \times m$ matrix, real or complex. This means there is a smallest integer r such that $I, A, \ldots, A^{r-1}$ are linearly independent, and hence there are real numbers $\alpha_{r-1}, \alpha_{r-2}, \ldots, \alpha_0$ such that

$$A^r + \alpha_{r-1}A^{r-1} + \cdots + \alpha_0 I = 0.$$

Defining

$$\phi_0(\lambda) = \lambda^r + \alpha_{r-1}\lambda^{r-1} + \cdots + \alpha_0, \tag{4.1}$$

we have that $\phi_0(A) = 0$. This monic polynomial (leading coefficient 1) is the monic polynomial of least degree for which $\phi_0(A) = 0$ and is called the <u>minimal polynomial of</u> A. The polynomial

$$\phi(\lambda) = \det\,(\lambda I - A) = \prod_{j=1}^{m} (\lambda - \lambda_j)$$

is called the <u>characteristic polynomial of</u> A, and the Hamilton-Cayley Theorem states that $\phi(A) = 0$ -- <u>every square matrix satisfies its characteristic equation</u>. Therefore, we know that the degree r of the minimal polynomial of A is less than or equal to m.

4.1. <u>Exercise</u>. Show that:

a. If ψ is a polynomial for which $\psi(A) = 0$, then the minimal polynomial of A divides ψ (i.e., $\psi(\lambda) = q(\lambda)\phi_0(\lambda)$ for some polynomial q).

b. Each root of the minimal polynomial of A is an eigenvalue of A, and each eigenvalue A is a root of the minimal polynomial.

We now let

$$\psi(\lambda) = \prod_{j=1}^{s} (\lambda-\lambda_j) = \lambda^s + a_{s-1}\lambda^{s-1} + \cdots + a_0$$

be any polynomial for which $\psi(A) = 0$. We need not assume, in so far as the algorithm is concerned, that the roots of ψ are eigenvalues of A, although we do know by Exercise 4.1 that each eigenvalue of A is a root of ψ. It would be advantageous for the computation of A^n to take ψ to be the minimal polynomial of A, but this may be difficult to compute. One can always take ψ to be the characteristic polynomial of A. Define relative to ψ

$$Q_j = (A-\lambda_j)Q_{j-1}, \quad Q_0 = I, \tag{4.2}$$

so that $Q_s = 0$, and

$$AQ_{j-1} = Q_j + \lambda_j Q_{j-1}, \quad j \geq 1.$$

Thus $A^0 = Q_0$, $A = AQ_0 = Q_1 + \lambda_1 Q_0$, $A^2 = Q_2 + (\lambda_1+\lambda_2)Q_1 + \lambda_1^2 Q_0, \ldots,$ and we see that

$$A^n = \sum_{j=1}^{s} w_j(n)Q_{j-1}, \tag{4.3}$$

where the $w_j(n)$ are to be determined. Letting $X(n) = \sum_{j=1}^{s} w_j(n)Q_{j-1}$, we need only find $w_j(n)$ for which $X(0) = I$ and $AX(n) = X(n+1)$. Since we have not assumed $\psi(\lambda)$ is the minimal polynomial of A, the Q_j may not be linearly independent and (4.3) need not determine the $w_j(n)$ uniquely. However, the initial condition is satisfied by selecting

$$w_1(0) = 1, \quad w_2(0) = \cdots = w_s(0) = 0,$$

and $X(n+1) = AX(n)$ is satisfied if

$$\sum_{j=1}^{s} w_j(n+1)Q_{j-1} = A(\sum_{j=1}^{s} w_j(n)Q_{j-1}) = \sum_{j=1}^{s} w_j(n)(Q_j+\lambda_j Q_{j-1}).$$

Hence, (4.3) holds if

$$\begin{aligned} w_1' &= \lambda_1 w_1, \quad w_1(0) = 1 \quad (w_1(n) = \lambda_1^n) \\ w_j' &= \lambda_j w_j + w_{j-1}, \quad w_j(0) = 0, \quad j = 2,\ldots,s. \end{aligned} \tag{4.4}$$

The equations (4.4) and (4.2) are algorithms for computing Q_j and $w_j(n)$; i.e., for computing A^n given the roots of the polynomial ψ. Such polynomials (for instance, the characteristic polynomial) can be computed. Note that the algorithm is valid for A any square matrix -- real or complex.

4.2. Example. To illustrate this algorithm let us find the solution of

$$y''' - 3y'' + 3y' - y = 0$$

satisfying $y''(0) = y_0''$, $y'(0) = y_0'$ and $y(0) = y_0$. This 3rd-order equation is equivalent to $x' = Ax$ where

$$x = \begin{pmatrix} x_1 \\ x_2 \\ x_3 \end{pmatrix} = \begin{pmatrix} y \\ y' \\ y'' \end{pmatrix} \quad \text{and} \quad A = \begin{pmatrix} 0 & 1 & 0 \\ 0 & 0 & 1 \\ 1 & -3 & 3 \end{pmatrix}.$$

Here we take ψ to be the characteristic polynomial

$$\phi(\lambda) = \det(A-\lambda I) = -(\lambda-1)^3 \quad \text{and} \quad \lambda_1 = \lambda_2 = \lambda_3 = 1.$$

$$Q_0 = I,\ Q_1 = A - I = \begin{pmatrix} -1 & 1 & 0 \\ 0 & -1 & 1 \\ 1 & -3 & 2 \end{pmatrix}$$

$$Q_2 = (A-I)^2 = \begin{pmatrix} 1 & -2 & 1 \\ 1 & -2 & 1 \\ 1 & -2 & 1 \end{pmatrix}.$$

Solving (4.4) directly, or by using Exercise 4.3, we obtain $w_1(n) = 1$, $w_2(n) = n$, $w_3(n) = \frac{1}{2}n(n-1)$. Hence

$$A^n = I + n(A-I) + \frac{1}{2}n(n-1)(A-I)^2$$

$$= \begin{pmatrix} \frac{1}{2}(n-1)(n-2) & -n(n-2) & \frac{1}{2}n(n-1) \\ \frac{1}{2}n(n-1) & -(n+1)(n-1) & \frac{1}{2}(n+1)n \\ \frac{1}{2}(n+1)n & -(n+2)n & \frac{1}{2}(n+2)(n+1) \end{pmatrix}.$$

The solution $y(n)$ is the first component of $A^n x^0$. This gives $y(n) = \frac{1}{2}(n-1)(n-2)y_0 - n(n-2)y_0' + \frac{1}{2}n(n-1)y_0''$.

4.3. <u>Exercise</u>. Show that the solution of (4.4) is

$$w_1(n) = \lambda_1^n$$

$$w_j(n+1) = \sum_{k=0}^{n} \lambda_j^{n-k} w_{j-1}(k), \qquad j = 2,\ldots,s.$$

4.4. <u>Exercise</u>. If the eigenvalues of $\lambda_1,\ldots,\lambda_s$ (the roots of ψ) are distinct, show that

$$w_1(n) = \lambda_1^n$$

$$w_j(n) = \sum_{i=1}^{j} c_{ij}\lambda_i^n, \qquad j = 2,\ldots,s,$$

where

$$c_{ij}^{-1} = \prod_{k=1,k\neq i}^{j} (\lambda_i - \lambda_k).$$

Rewriting equation (4.4), we have

$$w' = Bw \tag{4.5}$$

where

$$B = \begin{pmatrix} \lambda_1 & 0 & \cdots & & 0 \\ 1 & \ddots & \ddots & & \vdots \\ 0 & \ddots & \ddots & \ddots & 0 \\ \vdots & \ddots & \ddots & \ddots & \\ 0 & \cdots & 0 & 1 & \lambda_s \end{pmatrix},$$

and $w = (w_j)$ is the first principal solution of (3.5).

What we have done is to reduce the problem of computing A^n to that of computing B^n. In other words, the problem of solving $x' = Ax$ has been reduced to that of finding the first principal solution of (4.5). The next exercise shows that, in turn, this is equivalent to finding the last principal solution of the sth-order (scalar) equation $(y^{(s)}(n) = z^s y(n) = y(n+s))$

$$y(s) + a_{s-1}y^{(s-1)} + \cdots + a_0 y = 0 \tag{4.6}$$

or

$$\psi(z)y = (z-\lambda_1)(z-\lambda_2)\cdots(z-\lambda_s)y = 0.$$

Then

$$w = \begin{pmatrix} \psi_{s-1}(z)y \\ \vdots \\ \psi_2(z)y \\ \psi_1(z)y \\ y \end{pmatrix}, \tag{4.7}$$

where

$$\psi_1(\lambda) = \lambda - \lambda_s$$

$$\psi_j(\lambda) = (\lambda-\lambda_j)\psi_{j-1} = \prod_{i=j}^{s} (\lambda-\lambda_i), \quad j \geq 2,$$

and y is the last principal solution (4.6). As is explained in Section 10, B is a companion matrix of $\psi(\lambda)$ -- $\psi(\lambda)$ is both the minimal and characteristic polynomial of B and each eigenvalue λ_i of B has geometric multiplicity 1.

4.5. Exercise. Show that equation (4.5) is equivalent to (4.6) in the sense that, if w is a solution of (4.5), then $y = w_s$ is a solution of (4.6), and, conversely, if y is a solution of (4.6) then (4.7) is a solution of (4.5).

5. A characterization of stable matrices. Computational criteria.

We saw earlier in Section 3 that $r(A) < 1$ was a necessary condition for A to be stable. We now want to see that this condition is also sufficient and will prove this using the algorithm of the previous section. We give first an inductive and elementary proof, and then look at another proof that is more sophisticated and teaches us something about nonnegative matrices, which arise in and are important for many applications.

With reference to the algorithm, let us assume, as we always can, that $\psi(\lambda) = \prod_{j=1}^{s} (\lambda-\lambda_j)$ is the minimal polynomial of A. Then (Exercise 4.1)

$$\sigma(A) = \sigma(B) = \{\lambda_1,\ldots,\lambda_s\},$$

where

$$B = \begin{pmatrix} \lambda_1 & 0 & \cdots & & 0 \\ 1 & \ddots & \ddots & & \vdots \\ 0 & \ddots & \ddots & \ddots & 0 \\ \vdots & & \ddots & \ddots & \\ 0 & \cdots & 0 & 1 & \lambda_s \end{pmatrix}, \tag{5.1}$$

and A and B have the same spectral radius $(r(A) = r(B) = r_0)$. Using the w_i's of the algorithm, we see that

$$w(n) = \begin{bmatrix} w_1(n) \\ \vdots \\ w_2(n) \end{bmatrix}$$

is the first principal solution of (equation 4.5)

$$w' = Bw; \tag{5.2}$$

i.e., (equation 4.4)

$$\begin{aligned} w_1' &= \lambda_1 w_1, \quad w_1(0) = 1 \\ w_j' &= \lambda_j w_j + w_{j-1}, \quad w_j(0) = 0, \quad j = 2,\ldots,s. \end{aligned} \tag{5.3}$$

Then with $r_0 = r(A)$,

$$\begin{aligned} |w_1'| &\leq r_0 w_1 \\ |w_j'| &\leq r_0 |w_j| + |w_{j-1}|, \end{aligned} \tag{5.4}$$

and we want to find solutions of this difference inequality. It is easy enough to guess that, if $\beta > r_0$, then

$$|w_j(n)| \leq c_j \beta^n.$$

Substituting into (5.4) gives

$$c_j \leq \frac{c_{j-1}}{(\beta - r_0)}, \quad c_1 \geq 1.$$

Taking $c_1 = 1$, we obtain

$$c_j = \frac{1}{(\beta - r_0)^{j-1}}.$$

Substituting into (5.4), we see (by induction) that, if the w_j satisfy (5.3) then for all $n \geq 0$

$$|w_j(n)| \leq \frac{\beta^n}{(\beta - r_0)^{j-1}}, \quad j = 1,\ldots,s. \tag{5.5}$$

It then follows from equation (4.3) that, if $\beta > r(A)$, there is a constant γ such that

$$||A^n|| \leq \gamma\beta^n, \quad \text{all} \quad n \geq 0. \tag{5.6}$$

Hence we obtain

5.1. Theorem. A matrix A is stable if and only if $r(A) < 1$; i.e., if and only if all of its eigenvalues lie inside the unit circle.

There are many computational criteria for determining whether or not the roots of a polynomial $p(\lambda)$ lie inside the unit circle, where

$$p(\lambda) = a_m\lambda^m + a_{n-1}\lambda^{m-1} + \cdots + a_0, \quad a_m > 0. \tag{5.7}$$

We will give a statement of one such criterion -- usually called the Schur-Cohn criterion -- for the case where the a_i are real. We need first to explain what is meant by the "inners" of a square matrix.

The inners of a square matrix A are the matrix itself and all the matrices obtained by omitting successively the first and last rows and the first and last columns. For instance, for $m = 4$

$$\begin{pmatrix} a_{11} & a_{12} & a_{13} & a_{14} & a_{15} \\ a_{21} & \boxed{\begin{matrix} a_{22} & a_{23} & a_{24} \\ a_{32} & \boxed{a_{33}} & a_{34} \\ a_{42} & a_{43} & a_{44} \end{matrix}} & & & a_{25} \\ a_{31} & & & & a_{35} \\ a_{41} & & & & a_{45} \\ a_{51} & a_{52} & a_{53} & a_{54} & a_{55} \end{pmatrix}$$

and for $m = 5$

$$\begin{pmatrix} a_{11} & a_{12} & a_{13} & a_{14} & a_{15} & a_{16} \\ a_{21} & \boxed{a_{22}} & a_{23} & a_{24} & a_{25} & a_{26} \\ a_{31} & a_{32} & \boxed{a_{33}} & a_{34} & a_{35} & a_{36} \\ a_{41} & a_{42} & a_{43} & a_{44} & a_{45} & a_{46} \\ a_{51} & a_{52} & a_{53} & a_{54} & a_{55} & a_{56} \\ a_{61} & a_{62} & a_{63} & a_{64} & a_{65} & a_{66} \end{pmatrix}$$

A matrix is said to be positive <u>innerwise</u> if the determinants of all of its inners are positive.

5.2. <u>Exercise</u>. a. A real $m \times m$ matrix B is said to be <u>positive definite</u> if $x^T Bx = \sum_{i,j} b_{ij} x_i x_j > 0$ for $x \neq 0$. Show that B innerwise positive is a necessary condition that B (not necessarily symmetric) be positive definite.

5.3. <u>Proposition</u> (<u>Schur-Cohn Criterion</u>). A necessary and sufficient condition that the polynomial (5.7) with real coefficients have all of its roots inside the unit circle is that

(i) $p(1) > 0$ and $(-1)^m p(-1) > 0$

and

(ii) the $(m-1) \times (m-1)$ matrices

$$\Delta^{\pm}_{m-1} = \begin{pmatrix} a_m & 0 & \cdots & & 0 \\ a_{m-1} & a_m & & & \vdots \\ \vdots & \ddots & \ddots & & \vdots \\ a_3 & & \ddots & \ddots & 0 \\ a_2 & a_3 & \cdots & a_{m-1} & a_m \end{pmatrix} \pm \begin{pmatrix} 0 & \cdots & & 0 & a_0 \\ \vdots & & & a_0 & a_1 \\ \vdots & & \iddots & \iddots & \vdots \\ 0 & \iddots & \iddots & & a_{m-1} \\ a_0 & a_1 & \cdots & a_{m-1} & a_{m-2} \end{pmatrix}$$

are both positive innerwise.

5.4. Example. a. $m = 2$, $p(\lambda) = \lambda^2 + a_1\lambda + a_0$. $p(1) = 1 + a_1 + a_0$, $p(-1) = 1 - a_1 + a_0$ and $\Delta_1^{\pm} = 1 \pm a_0 > 0$. Thus, the roots lie inside the unit circle if and only if $|a_0| < 1$ and $|a_1| < 1 + a_0$.

b. $m = 3$, $p(\lambda) = \lambda^3 + a_2\lambda^2 + a_1\lambda + a_0$; $\psi(1) = 1 + a_2 + a_1 + a_0$, $-\psi(-1) = 1 - a_2 + a_1 - a_0$, and

$$\det(\Delta_2^{\pm}) = \begin{vmatrix} 1 & \pm a_0 \\ a_2 \pm a_0 & 1 \pm a_1 \end{vmatrix} > 0.$$

The roots lie inside the unit circle if and only if

$$|a_0 + a_2| < 1 + a_1 \quad \text{and} \quad |a_1 - a_0 a_2| < 1 - a_0^2.$$

5.5. Exercise. Show that the roots of the real polynomial $\lambda^4 + a_3\lambda^3 + a_2\lambda^2 + a_1\lambda + a_0$ all lie inside the unit circle if and only if

$$|a_0| < 1, \ |a_1 + a_3| < 1 + a_2 + a_0, \quad \text{and}$$

$$|a_2(1-a_0) + a_0(1-a_0^2) + a_3(a_0 a_3 - a_1)|$$
$$< a_0 a_2(1-a_0) + (1-a_0^2) + a_1(a_0 a_3 - a_1).$$

For other criteria and algorithms for computing the determinants of the inners of matrices see Jury [1].

5.6. Exercise. The trace of a matrix A is $\mathrm{Trace}(A) = \sum_{j=1}^{m} a_{jj}$. Establish the following necessary and sufficient conditions for a real $m \times m$ matrix A to be stable

a. $m = 2$: $|\det A| < 1$ and $|\mathrm{Trace}\ A| < 1 + \det A$.

b. $m = 3$: $|\det A + \mathrm{Trace}\ A| < 1 + a_1$

and

$$|a_1 - (\text{Trace } A)(\det A)| < 1 - (\det A)^2,$$

where

$$a_1 = \begin{vmatrix} a_{11} & a_{12} \\ a_{21} & a_{33} \end{vmatrix} + \begin{vmatrix} a_{11} & a_{13} \\ a_{31} & a_{33} \end{vmatrix} + \begin{vmatrix} a_{22} & a_{23} \\ a_{32} & a_{33} \end{vmatrix} .$$

We will now derive another inequality that implies (5.6). To do this we will consider nonnegative matrices and the concept of an absolute value of a matrix. For $B = (b_{ij})$ any real B <u>nonnegative</u>, written $B \geq 0$, means each $b_{ij} \geq 0$. The inequality $B \geq C$ (for comparable matrices) means $B-C \geq 0$. Thus, for $c \in \mathbf{R}^m$, $c = (c_i) \geq 0$ means each component $c_i \geq 0$. For any matrix B, the <u>absolute value</u> ($|B|$) of B is defined by $|B| = (|b_{ij}|)$. Thus,

$$|c| = \begin{pmatrix} |c_1| \\ \vdots \\ |c_m| \end{pmatrix} .$$

The basic elementary properties of nonnegative matrices, matrix inequalities, and the absolute value are given in Exercise 5.7.

5.7. <u>Exercise</u>. Show, wherever the operations are defined, that

a. $B \geq 0$ is equivalent to $Bc \geq 0$ for all $c \geq 0$ (B a real $r \times m$ matrix and $c \in \mathbf{R}^m$.

b. (i) $A \leq B$ and $B \leq C$ imply $A \leq C$.

(ii) $A \geq 0$ and $B \geq 0$ imply $A+B \geq 0$.

(iii) $B \leq C$ and $D \geq 0$ imply $DB \leq DC$ and $BD \leq CD$.

c. (i) $|B| \geq 0$, and $|B| = 0$ if and only if $B = 0$.

(ii) $|B+C| \leq |B| + |C|$.

(iii) $|\alpha B| \leq |\alpha|\,|B|$.

(iv) $|AB| \leq |A||B|$.

We will need in a moment the following relationship between the norm and the absolute value.

5.8. <u>Proposition</u>. Let A and B be $r \times m$ matrices. Then $||A|| \leq |||A|||$, and $0 \leq A \leq B$ implies $||A|| \leq ||B||$; i.e., $|A| \leq B$ implies $||A|| \leq |||A||| \leq ||B||$.

<u>Proof</u>: Note, for $c \in \mathbf{C}^m$ and $d \in \mathbf{R}^m$, that $|c| \leq d$ implies $||c|| = |||c||| \leq ||d||$. It follows from the definition of $||A||$ that there exist c and d such that $||c|| = ||d|| = 1$, $d \geq 0$, $||Ac|| = ||A||$, and $|||A|d|| = |||A|||$. Then

$$||A|| = ||Ac|| = |||Ac||| \leq |||A||c||| \leq |||A|||.$$

If $|A| \leq B$, then $|||A||| = ||\,A\,d|| \leq ||Bd|| \leq ||B||$. □

5.9. <u>Exercise</u>. Let $A = \begin{pmatrix} 1 & -1 \\ 1 & 2 \end{pmatrix}$. Show that $||A|| < |||A|||$. (One way to compute the norms is given in Exercise 3.5.)

Let us now return to a consideration of equation (5.2). We see that

$$|B| \leq B_0 = r_0 I + N,$$

where $r_0 = r(B) = r(A)$ and

$$N = \begin{pmatrix} 0 & \cdots & & & 0 \\ 1 & \ddots & & & \vdots \\ 0 & \ddots & \ddots & & \vdots \\ \vdots & \ddots & \ddots & \ddots & \\ 0 & \cdots & 0 & 1 & 0 \end{pmatrix}$$

Note that $N\begin{pmatrix} c_1 \\ \vdots \\ c_s \end{pmatrix} = \begin{pmatrix} 0 \\ c_1 \\ \vdots \\ c_{s-1} \end{pmatrix}$ and $N^s = 0$; i.e., N is nilpotent of order s. Then

$$|B|^n \leq B_0^n = (r_0 I+N)^n = \sum_{j=0}^{s-1} \binom{n}{j} r_0^{n-j} N^j,$$

where

$$\binom{n}{j} = \begin{cases} \dfrac{n!}{j!(n-j)!}, & n \geq j \geq 0 \\ 0, & \text{otherwise;} \end{cases}$$

the $\binom{n}{j}$ are the binomial coefficients. For any $\beta > r_0 = r(A) = r(B)$

$$\beta^{-n} B_0^n = \sum_{j=0}^{s-1} \beta^j \binom{n}{j} \left(\frac{r_0}{\beta}\right)^{n-j} N^j,$$

and we see that $\beta^{-n} B_0^n \to 0$ as $n \to \infty$. Hence, given $\beta > r(A)$ there is a matrix C such that

$$|B|^n \leq \beta^n C, \quad \text{all } n \geq 0. \tag{5.8}$$

Since $w(n)$ (equation 4.3)) is the first column of B^n ($|B^n| \leq |B|^n$), we have that

$$|w_i(n)| \leq \beta^n c_{i1}.$$

Then Equation (4.3) implies (for A any real or complex $(m \times m)$-matrix) that, if $\beta > r(A)$, there is a matrix K such that

$$|A^n| \leq \beta^n K, \quad \text{for all } n \geq 0. \tag{5.9}$$

Hence from Proposition 5.8 we obtain again the inequality (5.6) with $\gamma = ||K||$.

5.10. Exercise. Show that the inequality (5.8) holds with

$$c_{ij} = \begin{cases} 0 & , \quad \text{if} \quad i < j \\ \dfrac{1}{(\beta - r_0)^{i-j}} & , \quad \text{if} \quad i \geq j. \end{cases}$$

Note this implies (5.5).

6. Liapunov's characterization of stable matrices. A Liapunov function for $x' = Ax$.

Although Liapunov did not consider difference equations, what we do here is the exact analog of what Liapunov did for linear differential equations. In the context of differential equations a matrix is said to be stable if $e^{At} \to 0$ as $t \to \infty$, and for difference equations A^n is the analog of e^{At}.

Here we shall restrict ourselves to real matrices and consider the quadratic form

$$V(x) = x^T B x = \sum_{i,j=1}^{m} b_{ij} x_i x_j ,$$

where B is a real $m \times m$ matrix. A matrix B is said to be positive (negative) definite if $V(x)$ is positive (negative) definite, and, since

$$V(x) = \frac{1}{2} x^T (B+B^T) x,$$

B is positive definite if and only if $B + B^T$ is positive definite. We could, therefore, always restrict ourselves to B symmetric (i.e., $B = B^T$), and in practice this is what one would do. One reason for symmetrizing B is Sylvester's criterion which we now state.

6.1. Proposition (Sylvester's Criterion). A real symmetric

square matrix B is positive definite if and only if the determinants of its leading principal minors are positive; i.e., if and only if

$$b_{11} > 0, \quad \begin{vmatrix} b_{11} & b_{12} \\ b_{12} & b_{22} \end{vmatrix} > 0, \ldots, \det B > 0.$$

Note that this criterion does not hold if B has not been symmetrized. For example, $B = \begin{pmatrix} 1 & 0 \\ -4 & 1 \end{pmatrix}$. Another reason for making B symmetric in applications is that it reduces the number of unknowns in the equations to be dealt with.

The <u>minors</u> of a matrix B are the matrix itself and the matrix obtained by removing successively a row and a column. The <u>leading principal minors</u> are B itself and the minors obtained by removing successively the last row and the last column. The <u>principal minors</u> are B itself and the matrices obtained by removing successively an ith row and an ith column. For instance, for $n = 3$ the principal minors are

$$\begin{pmatrix} b_{11} & b_{12} & b_{13} \\ b_{21} & b_{22} & b_{23} \\ b_{31} & b_{32} & b_{33} \end{pmatrix}, \quad \begin{pmatrix} b_{11} & b_{12} \\ b_{21} & b_{22} \end{pmatrix}, \quad (b_{11})$$

$$\begin{pmatrix} b_{11} & b_{13} \\ b_{31} & b_{33} \end{pmatrix}, \quad \begin{pmatrix} b_{22} & b_{23} \\ b_{32} & b_{33} \end{pmatrix}, \quad (b_{22}), \ (b_{33}).$$

The first three of these are the leading principal minors. The matrix

$$\begin{pmatrix} b_{21} & b_{22} \\ b_{31} & b_{32} \end{pmatrix}$$

is a minor but not a principal minor.

It is convenient to note also that a <u>real symmetric matrix is positive definite if and only if the determinants of all the principal minors are positive</u>. This follows from Sylvester's criterion since the criterion does not depend on the numbering of the coordinates. It also follows from Sylvester's criterion that a necessary condition for a real symmetric matrix B to be positive semidefinite (B is <u>positive</u> (<u>negative</u>) <u>semidefinite</u> if $x^T Bx \geq 0$ (≤ 0) for all x.) is that the determinants of the leading principal minors be nonnegative. This is, however, not a sufficient condition. For instance,

$$\begin{pmatrix} 0 & 0 \\ 0 & -1 \end{pmatrix}$$

is not positive semidefinite. This necessary condition can be seen by considering $B + \varepsilon I$, $\varepsilon > 0$, and letting $\varepsilon \to 0$. This type of argument does give

6.2. <u>Proposition</u>. A real symmetric square matrix B is positive semidefinite if and only if the determinants of all of its principal minors are nonnegative.

We now apply the direct method of Liapunov to obtain another characterization of stable matrices. Returning to equation (3.1), $x' = Ax$, we take as a Liapunov function $V(x) = x^T Bx$, where B is positive definite, and $\dot{V}(x) = x^T(A^T BA-B)x$. If $A^T BA-B$ is negative definite, then (see Section 2) the origin for (3.1) is asymptotically stable, and A is stable. Conversely, suppose that A is stable, and consider the equation

$$A^T BA - B = -C, \qquad (6.1)$$

where C is a given matrix. If this equation has a solution, then

$$(A^T)^{k+1} BA^{k+1} - (A^T)^k BA^k = -(A^T)^k CA^k,$$

and summing gives

$$(A^T)^{n+1} BA^{n+1} - B = -\sum_{k=0}^{n} (A^T)^k CA^k.$$

Letting $n \to \infty$, we obtain

$$B = \sum_{k=0}^{\infty} (A^T)^k CA^k.$$

It is easy to verify that this is a solution of (6.1); and, if C is positive definite, then so is B, and, if C is symmetric, then so is B. Thus we have shown that

6.3. <u>Theorem</u>. If there are positive definite matrices B and C satisfying (6.1), then A is stable. Conversely, if A is stable, then, given any C, equation (6.1) has a unique solution; if C is positive definite, B is positive definite and, if C is symmetric B is symmetric.

This result yields a converse theorem for linear systems -- if the origin for $\dot{x} = Ax$ is asymptotically stable, then there is a quadratic function V satisfying Corollary

This converse result is useful in many ways and plays an important role in the theory of discrete linear control systems.

We now look at a corollary of Theorem 5.3 that we will find useful in the next section.

6.4. <u>Corollary</u>. If $r(A) \geq 1$ and equation (6.1) has a solution B for some positive definite matrix C, then B is either negative semidefinite or indefinite; i.e., x^TBx is negative for some x.

Proof: If $r(A) \geq 1$, then A is not stable, and B cannot be positive definite. But also B cannot be positive semi-definite. If it were then, for some $x \neq 0$, $x^TBx = 0$ and $x^TA^TBAx = -x^TCx < 0$, which is a contradiction. Hence B is either indefinite or negative semi-definite.

Let us consider the more general question of when equation (6.1) has a unique solution. To do this we consider the more general equation

$$A_1XA_2 - X = C, \tag{6.2}$$

where A_1 is an $m \times m$ matrix, A_2 is an $n \times n$ matrix and X and C are $m \times n$ matrices.

6.5. Proposition. Equation (6.2) has a unique solution if and only if no eigenvalue of A_1 is a reciprocal of an eigenvalue of A_2.

Proof: What we want to show is that this condition on A_1 and A_2 is equivalent to $A_1XA_2 = X$ implies $X = 0$. Assume first that the condition on A_1 and A_2 is satisfied. Now $A_1XA_2 = X$ implies $A_1^{k-j}XA_2^{k-j} = X$ and

$$A_1^jX = A_1^kXA_2^{k-j} \quad \text{for } k \geq j \geq 0.$$

Defining for a polynomial

$$p(\lambda) = \sum_{j=0}^{k} a_j\lambda^j$$

of degree k

$$p^*(\lambda) = \sum_{j=0}^{k} a_j\lambda^{k-j} = \lambda^k p(1/\lambda),$$

we see that

$$p(A)X = A_1^k X p^*(A_2).$$

Let $\phi_i(\lambda)$ be the characteristic polynomial of A_i. Then since $\phi_1(\lambda)$ and $\phi_2^*(\lambda)$ are relatively prime there are polynomials $p(\lambda)$ and $q(\lambda)$ such that

$$p(\lambda)\phi_1(\lambda) + q(\lambda)\phi_2^*(\lambda) = 1.$$

Define $\phi(\lambda) = q(\lambda)\phi_2^*(\lambda)$ and note that $\phi^*(\lambda) = q^*(\lambda)\phi_2(\lambda)$. Hence $\phi^*(A_2) = 0$ and $\phi(A_1) = I$. From this we see that $A_1XA_2 = X$ implies $X = 0$.

To prove the converse assume that λ is an eigenvalue of A_1 and λ^{-1} is an eigenvalue of A_2 (and hence is also an eigenvalue of A_2^T). Let $A_1x^1 = \lambda x^1$ and $A^Tx^2 = \lambda^{-1}x^2$, $x_1 \neq 0$ and $x_2 \neq 0$. Define $X = (x_1^2x^1, x_2^2x^1, \ldots, x_n^2x^1)$. Then $X \neq 0$ and $A_1XA_2 = X$. $\square$

If A is stable, then no reciprocal of an eigenvalue of A is an eigenvalue, and this proposition gives another way of showing that equation (6.2) has a unique solution B for each C if A is stable.

6.6. Exercise. Show that $A_1X - XA_2 = C$ has a unique solution if and only if A_1 and A_2 have no eigenvalues in common.

7. Stability by the linear approximation.

The oldest method of investigating the stability of a system is to replace the system by its linear approximation. This method was used long before Liapunov, although Liapunov appears to be the first person who justified the method for ordinary differential equations. In 1929 Perron [1] investigated the question of when the stability of the difference equation

$$x' = Ax \tag{7.1}$$

determines the stability of the nonlinear equation,

$$x' = Ax + f(x) \tag{7.2}$$

Here we follow Liapunov rather than Perron. We shall assume that $f(x)$ is $o(x)$ ("little oh of x") as norm of x goes to zero; i.e., given $\varepsilon > 0$ there is a $\delta > 0$ such that $|f(x)| < \varepsilon|x|$ for all $n \geq 0$ and all $|x| < \delta$. Near the origin we expect to be able to neglect the nonlinear terms $f(x)$. We now show in the next theorem that this is valid except in the critical case $r(A) = 1$.

7.1. Theorem. Assume that $f(x)$ is $o(x)$ as $||x|| \to 0$. If A is stable, then the origin is an asymptotically stable equilibrium point of (7.2). If $r(A) > 1$, then the origin is an unstable equilibrium point of (7.2).

Proof: Assume that A is stable. Then, for any positive definite matrix C, equation (6.1) has a positive definite solution B. As a matter of convenience, take $C = I$. Then with $V(x) = x^TBx$ we have relative to equation (7.2) that

$$\dot{V}(x) = -x^Tx + 2x^TBf(x) + V(f(x)).$$

For any $0 < \alpha < 1$ we can select δ sufficiently small that

$$\dot{V}(x) \leq -\alpha x^T x \quad \text{for all} \quad |x| < \delta.$$

Hence by Corollary 2.7 the origin is an asymptotically stable equilibrium point of (7.2).

Now let $r(A) > 1$, and select $\beta > 0$ so that no eigenvalue of $A_0 = (1+\beta)^{-1/2}A$ is an eigenvalue of A and so small that $r(A_0) > 1$ $((1+\beta)^{1/2} < r(A))$. Then by Proposition 6.5 and Corollary 6.4 there is a B satisfying

$$A_0^T B A_0 - B = -I \quad \text{or} \quad A^T BA - B = \beta B - (1+\beta)I,$$

where $V(x) = -x^T Bx$ takes on positive values for some x. Then again for any $0 < \alpha < 1$ and δ sufficiently small we have relative to (7.2) that

$$\dot{V}(x) = \beta V(x) + W(x)$$

where $W(x) \geq \alpha x^T x$ for all $|x| < \delta$. By Proposition 2.9 we see for (7.2) that the origin is uniformly unstable.

Determining stability by the linear approximation has a long history and was, and still is, important for many applications. Up until 1950, or there abouts, it was the only mathematical approach to the design and analysis of control systems and feedback devices, and this goes back to Maxwell [1] in 1868 and beyond.

This method of using the linear approximation to study stability has serious practical disadvantages since it does not take into account the nonlinearities of the system, and one of the advantages of Liapunov's direct method is that it does. A good first thing to do, however, is to take a look at the linear approximations. We will return later (in Section) to a discussion of these questions.

8. The general solution of $x' = Ax$. The Jordan Canonical Form.

We already know from the algorithm for computing A^n a good deal about the solutions of the linear homogeneous equation

$$x' = Ax. \tag{8.1}$$

However, the following result from linear algebra tells us a bit more.

8.1. Proposition (Jordan Canonical Form). Corresponding to each $m \times m$ matrix A there is a nonsingular matrix P such that

$$\hat{A} = P^{-1}AP = \operatorname{diag}\{Q_1(\lambda_1),\ldots,Q_s(\lambda_s)\} = \begin{pmatrix} Q_1(\lambda_1) & 0 & \cdots & 0 \\ 0 & Q_2(\lambda_2) & \ddots & \vdots \\ \vdots & \ddots & \ddots & 0 \\ 0 & \cdots & 0 & Q_s(\lambda_s) \end{pmatrix}$$

where each $Q_i(\lambda_i)$ is an $s_i \times s_i$ matrix of the form

$$Q_i = \begin{pmatrix} \lambda_i & 1 & 0 & \cdots & 0 \\ 0 & \ddots & \ddots & \ddots & \vdots \\ \vdots & & \ddots & \ddots & 0 \\ \vdots & & & \ddots & 1 \\ 0 & \cdots & & 0 & \lambda_i \end{pmatrix}$$

For instance, up to changes in the numbering of the eigenvalues, the possible Jordan canonical forms for $m = 3$ ($s = 3$, $s = 2$, and $s = 1$) are:

$$\text{(i)} \begin{bmatrix} \lambda_1 & 0 & 0 \\ 0 & \lambda_2 & 0 \\ 0 & 0 & \lambda_3 \end{bmatrix}, \quad \text{(ii)} \begin{bmatrix} \lambda_1 & 1 & 0 \\ 0 & \lambda_1 & 0 \\ 0 & 0 & \lambda_2 \end{bmatrix} \quad (\lambda_3 = \lambda_1), \text{ and}$$

$$\text{(iii)} \begin{bmatrix} \lambda_1 & 1 & 0 \\ 0 & \lambda_1 & 1 \\ 0 & 0 & \lambda_1 \end{bmatrix} \qquad (\lambda_3 = \lambda_2 = \lambda_1).$$

The polynomials $\phi_i(\lambda) = \det(\lambda I - Q_i(\lambda_i)) = (\lambda-\lambda_i)^{s_i}$, $i = 1,\ldots,s$, are called the <u>elementary divisors</u> of A. For $m = 3$ the elementary divisors are for the above 3 cases: (i) $(\lambda-\lambda_1)$, $(\lambda-\lambda_2)$, $(\lambda-\lambda_3)$; (ii) $(\lambda-\lambda_1)^2$, $(\lambda-\lambda_2)$; and (iii) $(\lambda-\lambda_1)^3$.

Two square matrices A and B are said to be <u>similar</u> if there is a nonsingular matrix P such that $B = P^{-1}AP$. Similar matrices represent, relative to different coordinate systems, the same linear transformation. We will return to a discussion of the Jordan canonical form and similar matrices in Section 10. Suffice it to say here that <u>two matrices are similar if and only if they have the same Jordan canonical form</u>, or equivalently, <u>the same elementary divisors</u>. The elementary divisors of a matrix are a complete set of invariants for the matrix.

From the point of view of our difference equation (8.1) Proposition 8.1 tells us that by a change of coordinates $\hat{x} = Px$, P nonsingular, we can decouple the system;

$$P = (v^1,\ldots,v^m) \quad \text{where} \quad v^1,\ldots,v^m$$

is the basis for the canonical coordinates $\hat{x}$. In the new coordinates $\hat{x}$ equation (8.1) becomes $(\hat{A} = P^{-1}AP)$

$$x' = \hat{A}\,\hat{x} \tag{8.2}$$

or

$$\hat{x}^{j\,\prime} = Q_j(\lambda_j)\hat{x}^j, \quad j = 1,\ldots,s, \quad \hat{x} = \begin{pmatrix} \hat{x}^1 \\ \cdot \\ \cdot \\ \cdot \\ \hat{x}^j \end{pmatrix} \tag{8.3}$$

Each equation in (8.3) is of the form

$$y' = Qy = \lambda_0 y + Ny, \tag{8.4}$$

where

$$N = (Q-\lambda_0 I) = \begin{bmatrix} 0 & 1 & 0 & \cdots & 0 \\ \cdot & \cdot & \cdot & \cdot & \vdots \\ \cdot & & \cdot & \cdot \;\; \cdot & \vdots \\ \cdot & & & \cdot \;\; \cdot & 0 \\ & & & \cdot & 1 \\ 0 & \cdot & \cdot & \cdot & 0 \end{bmatrix}$$

is an $r \times r$ matrix.

Note that

$$N\begin{bmatrix} y_1 \\ \vdots \\ y_r \end{bmatrix} = \begin{bmatrix} y_2 \\ \vdots \\ y_r \\ 0 \end{bmatrix} \quad \text{and} \quad N^r = 0;$$

i.e., N is nilpotent of order r. (A matrix B is said to be <u>nilpotent</u> if $B^r = 0$; if $B^{r-1} \neq 0$, r is the <u>order</u> of the nilpotence.)

All that we need to do to solve (8.4) is to compute Q^n and that is easy:

$$Q^n = (N+\lambda_0 I)^n = \sum_{j=0}^{n} \binom{n}{j} \lambda_0^{n-j} N^j$$

and

$$Q^n = \sum_{j=0}^{r-1} \binom{n}{j} \lambda_0^{n-j} N^j, \tag{8.5}$$

where the $\binom{n}{j}$ are the binomial coefficients --

$$\binom{n}{j} = \begin{cases} \dfrac{n!}{j!(n-j)!}, & n \geq j \geq 0; \\ 0, & \text{otherwise.} \end{cases}$$

Each $\binom{n}{j}$, $n \geq j \geq 0$, is a polynomial of degree j in n. Thus

$$Q^n = \begin{pmatrix} \lambda_0^n & n\lambda_0^{n-1} & \cdots & \binom{n}{r-1}\lambda_0^{n-r} \\ 0 & \lambda_0^n & \ddots & \vdots \\ \vdots & \ddots & \ddots & n\lambda_0^{n-1} \\ 0 & \cdots & 0 & \lambda_0^n \end{pmatrix},$$

and the general solution of (7.4) is

$$y(n) = \sum_{j=0}^{r-1} \binom{n}{j}\lambda_0^{n-j}N^j c, \tag{8.6}$$

where $y(0) = c$.

Going back to equation (8.2), we have

$$A^n = \mathrm{diag}(Q_1^n,\ldots,Q_s^n) = (\hat{x}^1(n),\ldots,\hat{x}^m(0)),$$

so that $\hat{x}^j(n)$ is the jth principal solution of (8.2) and

$$\hat{x}^1(n) = \begin{pmatrix} \lambda_1^n \\ 0 \\ \vdots \\ 0 \end{pmatrix}, \quad \hat{x}^2(n) = \begin{pmatrix} n\lambda_1^{n-1} \\ \lambda_1^n \\ 0 \\ \vdots \\ 0 \end{pmatrix}, \ldots, \hat{x}^{s_1} = \begin{pmatrix} \binom{n}{s_1-1}\lambda_1^{n-s_1+1} \\ \vdots \\ n\lambda_1^{n-1} \\ \lambda_1^n \end{pmatrix},$$

$$\hat{x}^{s_1+1}(n) = \begin{pmatrix} \lambda_2^n \\ 0 \\ \vdots \\ 0 \end{pmatrix}, \quad \cdots .$$

Letting $P = (v^1,\ldots,v^m)$, the functions

$$\xi^j(n) = P\hat{x}^j(n) = \sum_{j=1}^{m} \hat{x}_i^j v^i, \quad j = 1,\ldots,m, \tag{8.7}$$

is a basis of solutions of (8.1). To compute the general solution of (8.1) this way requires computing the eigenvalues of A and the vectors v^i (see Exercise 8.3). This is not usually a good computational procedure, and our main interest here is in the general information the Jordan canonical form gives us about the nature of the solutions of (8.1). For instance, this gives us another proof that A^n is stable if and only if $r(A) < 1$. Note also, we have shown

8.2. Proposition. The component $x_i(n)$ of each solution $x(n)$ of (8.1) is a linear combination of $\binom{n}{j}\lambda_i^{n-j}$, $j = 0,\ldots,s_i-1$, $i = 1,\ldots,s$.

8.3. Exercise. Show that:

$$
\begin{aligned}
(A-\lambda_1 I)v^1 &= 0\\
(A-\lambda_1 I)v^2 &= v\\
&\vdots\\
(A-\lambda_1 I)v^{s_1} &= v^{s_1-1}\\
(A-\lambda_2)v^{s_1+1} &= 0\\
&\vdots
\end{aligned}
$$

and hence

$$
\begin{aligned}
(A-\lambda_1)^{s_1}v^{s_1} &= 0\\
v^{s_1-1} &= (A-\lambda_i I)v^{s_1}\\
&\vdots\\
v^1 &= (A-\lambda_i I)^{s_1-1}v^{s_1} \neq 0\\
(A-\lambda_2 I)^{s_2}v^{s_1+s_2} &= 0\\
&\vdots
\end{aligned}
$$

Discuss a procedure for determining the v^i's given the eigenvalues of A (i.e., given $\sigma(A)$).

8.4. Exercise. A square matrix S is said to be diagonalizable (or semisimple) if S is similar to a diagonal matrix (the algebraic multiplicity of each eigenvalue is equal to its geometric multiplicity 1. Show that: Each square matrix A is uniquely expressible in the form $A = S + N$, where S is semisimple, $\sigma(S) = \sigma(A)$, N is nilpotent and $NS = SN$. What is the order of the nilpotence of S?

9. Higher order equations. The general solution of $\psi(z)y = 0$.

We want now to examine the general solution of the mth-order difference equation

$$\psi(z)y = y^{(m)} + a_{m-1}y^{(m-1)} + \cdots + a_0 y = 0; \tag{9.1}$$

$$\psi(\lambda) = \lambda^m + a_{m-1}\lambda^{m-1} + \cdots + a_0 = (\lambda-\lambda_1)(\lambda-\lambda_2)\cdots(\lambda-\lambda_m) = 0$$

is called the characteristic equation of (9.1).

A direct way to solve (9.1) is to look for solutions of the form $y(n) = \lambda^n$. Substituting gives

$$\psi(z)\lambda^n = \lambda^n\psi(\lambda) = 0,$$

so that λ^n is a solution if and only if λ is a root of $\psi(\lambda)$. Therefore

$$y(n) = c_1\lambda_1^n + c_2\lambda_2^n + \cdots + c_n\lambda_2^n \tag{9.2}$$

is a solution for all c_i. If the roots λ_i are distinct, then we can see this is the general solution of (9.1). Every solution is uniquely determined by its initial values

$$y(0) = b_i,\quad y(1) = b_2,\ldots,y(m-1) = b_m,$$

and every solution will be of the form (9.2) if for arbitrary b_i there are solutions c_i to the equations

$$\begin{aligned} c_1 + c_2 + \cdots + c_m &= b_1 \\ c_1\lambda_1 + c_2\lambda_2 + \cdots + c_m\lambda_m &= b_2 \\ &\vdots \\ c_1\lambda_1^{m-1} + c_2\lambda_2^{m-1} + \cdots + c_m\lambda_m^{m-1} &= b_m. \end{aligned}$$

where (9.2) is the general solution if

$$\Delta_m(\lambda_1,\ldots,\lambda_m) = \begin{vmatrix} 1 & 1 & \cdots & 1 \\ \lambda_1 & \lambda_2 & \cdots & \lambda_m \\ \vdots & \vdots & & \vdots \\ \lambda_1^{m-1} & \lambda_2^{m-1} & & \lambda_m^{m-1} \end{vmatrix} \neq 0;$$

$\Delta_m(\lambda_1,\ldots,\lambda_m)$ is called the <u>vandermonde</u> <u>determinant</u> (see Bellman [1], p. 186). Since Δ_m is a polynomial in the λ_i, it is not difficult to see that

$$\Delta_m(\lambda_1,\ldots,\lambda_m) = \prod_{1\le i\le j\le m} (\lambda_j - \lambda_i).$$

Therefore (9.2) is the general solution of (9.1) if and only if the λ_i are distinct.

9.1. <u>Exercise</u>. Solve $y'' - \sqrt{2}\, y' + 1 = 0$. The characteristic equation is $\lambda^2 - \sqrt{2}\,\lambda + 1$, and the roots are $\lambda_1 = \frac{1}{\sqrt{2}}(1+i)$, $\lambda_2 = \frac{1}{\sqrt{2}}(1-i)$. The general solution is, since $\lambda_1 = e^{i\,\pi/4}$ and $\lambda_2 = e^{-i\,\pi/4}$,

$$y(n) = c_1 e^{i\frac{n\pi}{4}} + c_2 e^{-i\frac{n\pi}{4}}.$$

The solutions are periodic with period 8. For real solutions $c_2 = \bar{c}_1$, and we see that all real solutions are of the form

$$y(n) = a \cos\left(\frac{n\pi}{2} + \delta\right)$$

or

$$y(n) = a_1 \cos\frac{n\pi}{2} + a_2 \sin\frac{n\pi}{2}\ .$$

In the case of multiple roots we need to find more solutions. At this point we could certainly guess what they are. Let us, however, find these additional solutions by replacing (9.1) by an equivalent system of first order equations. Letting

$$x = \begin{pmatrix} y \\ zy \\ \vdots \\ z^{m-1}y \end{pmatrix} = \begin{pmatrix} y \\ y' \\ \vdots \\ y^{(m-1)} \end{pmatrix},$$

equation (9.1) is equivalent to

$$x' = A_0 x \tag{9.3}$$

where

$$A_0 = \begin{pmatrix} 0 & 1 & 0 & \cdots & 0 \\ 0 & 0 & 1 & \ddots & \vdots \\ \vdots & \ddots & \ddots & \ddots & 0 \\ 0 & \cdots & 0 & 0 & 1 \\ -a_0 & -a_1 & & \cdots & -a_{m-1} \end{pmatrix}.$$

If $x(n)$ is any solution of (9.3), then $y(n) = x_1(n)$ is a solution of (9.1); and, conversely if $y(n)$ is a solution of (9.1)

$$x(n) = \begin{pmatrix} y(n) \\ y'(n) \\ \vdots \\ y^{(m-1)}(n) \end{pmatrix}$$

is a solution of (9.3).

We call A_0 the <u>principal companion matrix</u> of the polynomial $\psi(\lambda) = \lambda^m + a_{m-1}\lambda^{m-1} + \cdots a_0$. It is easy to see (by induction, for example) that $\det(\lambda I - A_0) = \psi(\lambda)$ -- the characteristic polynomial of A_0 is $\psi(\lambda)$. (We study companion matrices in more detail in the next section and give a more elegant proof. By removing the first column and the last row of $A_0 - \lambda I$, we see that the rank $(A_0 - \lambda I) \geq m-1$. For $\lambda \in \sigma(A_0)$, rank $(A_0 - \lambda I) < m$, and hence rank $(A_0 - \lambda I) = m-1$, $\lambda \in \sigma(A_0)$. This means that the equation $(A_0 - \lambda I)x = 0$ has one and only one linearly independent solution for each $\lambda \in \sigma(A_0)$ -- the geometric multiplicity of each eigenvalue is one. Therefore there is only one Jordan block $Q_j(\lambda_j)$ in the Jordan canonical form for each λ_j, and the minimal polynomial of A_0 is its characteristic polynomial. For example, the only possible Jordan canonical forms (once the eigenvalues have been numbered) for $m = 3$ are

$$\begin{pmatrix} \lambda_1 & 0 & 0 \\ 0 & \lambda_2 & \\ 0 & 0 & \lambda_3 \end{pmatrix}, \quad (\lambda_3 - \lambda_2)(\lambda_3 - \lambda_1)(\lambda_3 - \lambda_2) \neq 0;$$

$$\begin{pmatrix} \lambda_1 & 1 & 0 \\ 0 & \lambda_1 & 0 \\ 0 & 0 & \lambda_3 \end{pmatrix}, \quad \lambda_1 \neq \lambda_3; \text{ and}$$

$$\begin{pmatrix} \lambda_1 & 1 & 0 \\ 0 & \lambda_1 & 1 \\ 0 & 0 & \lambda_1 \end{pmatrix}.$$

Because of the equivalence of (9.1) and (9.3), we know from Proposition 8.2 that every solution $y(n) = x_1(n)$ of

(9.1) is a linear combination of the m functions

$$\binom{n}{j}\lambda_i^{n-j}, \quad j = 0,\ldots,s_{i-1}, \quad i = 1,\ldots,s; \tag{9.4}$$

$\lambda_1,\ldots,\lambda_s$ are the distinct roots of $\psi(\lambda)$, and the algebraic multiplicity of each λ_i is s_i (each λ_i is a root of multiplicity s_i of $\psi(\lambda)$). In (9.4) we can always replace $\binom{n}{j}$ by a polynomial $n(n-1)\cdots(n-j+1)$ of degree j, and, if $\lambda_i \neq 0$, we can replace $\lambda_i^{n-j} = \lambda_i^n$. Since

$$(z-\lambda_i)\binom{n}{j}\lambda_i^{n-j} = \binom{n}{j-1}\lambda_i^{n+1-j},$$

$$(z-\lambda_i)^k\binom{n}{j}\lambda_i^{n-j} = \binom{n}{j-k}\lambda_i^{n+k-j},$$

and

$$(z-\lambda_i)^{s_i}\binom{n}{j}\lambda_i^{n-j} = 0,$$

and each $\binom{n}{j}\lambda_i^{n-j}$ is a solution of (9.1). The general solution is a linear combination of these m solutions, and therefore span the space of all solutions of (9.1). Since this space has dimension m (Proposition 3.9), we have the following result.

9.2. <u>Theorem</u>. Let $\psi(\lambda) = \lambda^m + a_{m-1}\lambda^{m-1} + \cdots + a_0 = (\lambda-\lambda_1)^{s_1}(\lambda-\lambda_2)^{s_2}\ldots(\lambda-\lambda_s)^{s_s}$, where $\lambda_i \neq \lambda_j$ for $i \neq j$. Then the m solutions $\binom{n}{j}\lambda_i^{n-j}$, $j = 0,\ldots,s_i-1$, $i = 1,\ldots,s$ are a basis of solutions of

$$\psi(z)y = y^{(m)} + a_{m-1}y^{(m-1)} + \cdots + a_0 y = 0.$$

An interesting special case is $(\Delta = z-1)$

$$\Delta^{m+1}y = 0; \tag{9.5}$$

$\psi(\lambda) = (\lambda-1)^{m+1}$. Thus $1, n, \ldots, \binom{n}{m}$ is a basis of solutions of (9.5), and the general solution is

$$y(n) = c_0 + nc_1 + \cdots + \binom{n}{m} c_m = \sum_{j=0}^{m} \binom{n}{j} c_j .$$

It is clear that any polynomial of degree m is a solution of (9.5), and we have the following proposition as an immediate consequence of Theorem 9.2.

9.3. <u>Proposition</u>. Let $p(\lambda)$ be a polynomial of degree m. Then $p(\lambda)$ is uniquely expressible in the form

$$p(\lambda) = \sum_{j=0}^{m} \binom{\lambda}{j} c_j ,$$

where

$$\binom{\lambda}{j} = \frac{\lambda(\lambda-1)\ldots(n-j+1)}{j!}$$

<u>Proof</u>: We know that $p(\lambda)$ is a solution of (9.5), and Theorem 9.2 tells us the result is true for all $\lambda = n, n \geq 0$, and hence it is true for all λ. $\square$

9.4. <u>Example</u>. (Compare with Example 4.2). Let us determine the principal solutions of

$$y''' - 3y'' + 3y' - y = (z-1)^3 y = 0.$$

The general solution is

$$y(n) = c_1 + nc_2 + \frac{n(n-1)}{2} c_3 .$$

$$y(0) = c_1$$

$$y'(0) = c_1 + c_2$$

$$y''(0) = c_1 + 2c_2 + c_3 .$$

Solving

$$\begin{aligned} c_1 &= 1 \\ c_1 + c_2 &= 0 \\ c_1 + 2c_2 + c_3 &= 0, \end{aligned} \quad \text{gives} \quad \begin{aligned} c_1 &= 1 \\ c_2 &= -1 \\ c_3 &= 1. \end{aligned}$$

and the first principal solution is

$$y_1(n) = \frac{1}{2}(n-1)(n-2).$$

Similarly,

$$y_2(n) = -n(n-2)$$

$$y_3(n) = \frac{1}{2}\, n(n-1).$$

9.5. Exercise. a. Solve $\Delta^s y = (z-1)^s y = 0$ directly without using Theorem 9.2.

b. Show that the substitution $y(n) = \lambda^n \hat{y}(n)$ reduces $(z-\lambda)^s y = 0$ to $\Delta^s \hat{y} = 0$, and use this to prove Theorem 9.2.

10. Companion matrices. The equivalence of $x' = Ax$ and $\psi(z)y = 0$.

In some sense this section is a digression and could be postponed. However, the question we will ask is at this point a natural one, and its answer and the concept of companion matrices are general interest. What we do here is of special interest within the theory of the control and stability of continuous, as well as discrete, systems. We will see this for discrete systems in Section 15. The reader can, if he wishes, simply skim through this section and move on, and then come back later to pick up what is needed.

We look first at the case $m = 3$, which makes the generalization to arbitrary dimension easy. Let

$$f(\lambda) = \lambda^3 + a_2\lambda^2 + a_1\lambda + a_0, \tag{10.1}$$

and consider the 3rd-order difference equation

$$\psi(z)y = y''' + a_2 y'' + a_1 y' + a_0 y = 0. \tag{10.2}$$

Given $y(0)$, $y'(0) = y(1)$, and $y''(0) = y(2)$, equation (10.2) has a unique solution $y(n)$ for $n \geq 0$; $y(0)$, $y'(0)$, $y''(0)$ is the state of (10.2) at time 2 and $y(n)$, $y'(n)$, $y''(n)$ is the state of (10.2) at time $n + 2$. Letting

$$\overline{x} = \begin{bmatrix} y \\ y' \\ y'' \end{bmatrix},$$

We see that (10.2) is equivalent to the system

$$\begin{aligned} \overline{x}_1' &= \overline{x}_2 \\ \overline{x}_2' &= \overline{x}_3 \\ \overline{x}_3' &= -a_0\overline{x}_1 - a_1\overline{x}_2 - a_2\overline{x}_3, \end{aligned}$$

or

$$\overline{x}' = A_0\overline{x}, \tag{10.3}$$

where

$$A_0 = \begin{bmatrix} 0 & 1 & 0 \\ 0 & 0 & 1 \\ -a_0 & -a_1 & -a_2 \end{bmatrix}. \tag{10.4}$$

If $y(n)$ is a solution of (10.2), then $\overline{x}(n)$ is a solution of (10.3); conversely, if $\overline{x}(n)$ is a solution of (10.3), then $y(n) = \overline{x}_1(n)$ is a solution of (10.2). The matrix A_0 is usually called <u>the companion matrix</u> of $\psi(\lambda)$. We will call it <u>the principal companion matrix</u> of $\psi(\lambda)$. Any matrix A similar to A_0 we will call <u>a companion matrix</u> of $\psi(\lambda)$. Note that, if A is a companion of matrix of $\psi(\lambda)$, then the characteristic polynomial of A is $\psi(\lambda)$, since $\det(\lambda I - A_0) = \psi(\lambda)$.

There are other ways to reduce (10.2) to a system.

Let

$$x_1 = y$$
$$x_2 = (z-\lambda_1)y = y' - \lambda_1 y$$
$$x_3 = (z-\lambda_2)(z-\lambda_1)y = y'' - (\lambda_1+\lambda_2)y' + \lambda_1\lambda_2 y.$$

Then

$$x_1' = zy' = (z-\lambda_1 y) + \lambda_1 y$$

$$x_2' = z(z-\lambda_1)y = (z-\lambda_2)(z-\lambda_1)y + \lambda_2(z-\lambda_1)y$$

$$x_3' = z(z-\lambda_2)x_2 = (z-\lambda_1)(z-\lambda_2)(z-\lambda_3)y + \lambda_3(z-\lambda_1)(z-\lambda_2)y,$$

or

$$\begin{aligned} x_1' &= \lambda_1 x_1 + x_2 \\ x_2' &= \lambda_2 x_2 + x_3 \\ x_3' &= \lambda_3 x_3. \end{aligned} \tag{10.5}$$

Thus with

$$x = \begin{pmatrix} 1 & 0 & 0 \\ -\lambda_1 & 1 & 0 \\ \lambda_1\lambda_2 & -\lambda_1-\lambda_2 & 1 \end{pmatrix}\begin{pmatrix} y \\ y' \\ y'' \end{pmatrix} = P_1\hat{x},$$

(10.2) is equivalent to

$$x' = A_1 x, \quad A_1 = \begin{pmatrix} \lambda_1 & 1 & 0 \\ 0 & \lambda_2 & 1 \\ 0 & 0 & \lambda_3 \end{pmatrix}; \tag{10.6}$$

A_1 is also a companion matrix of $\psi(\lambda)$. Starting at the bottom the equations (10.5) can be solved recursively for $x_1(n) = y(n)$. What interests us here is that, we now see that $A_1 = P_1 A_0 P_1^{-1}$ -- A_1 is similar to A_0 -- and A_1 is a <u>companion matrix</u> of $\psi(z)$.

Another system equivalent to (10.2) is obtained by letting

$$x_1 = a_1 y + a_2 y' + y'' = a_1\bar{x}_1 + a_2\bar{x}_2 + \bar{x}_3$$

$$x_2 = a_2 y + y' \qquad\qquad = a_2\bar{x}_1 + \bar{x}_2$$

$$x_3 = y \qquad\qquad\qquad = \bar{x}_1 .$$

Then

$$x_3' = x_2 - ax_3$$

$$x_2' = x_1 - a_1 x_3$$

$$x_1' = -a_0 x_3;$$

i.e. with

$$x = P_2\bar{x}, \quad P_2 = \begin{bmatrix} a_1 & a_2 & 1 \\ a_2 & 1 & 0 \\ 1 & 0 & 0 \end{bmatrix},$$

(10.2) is equivalent to

$$x' = A_2 x, \quad A_2 = \begin{bmatrix} 0 & 0 & -a_0 \\ 1 & 0 & -a_1 \\ 0 & 1 & -a_2 \end{bmatrix} \tag{10.7}$$

and $A_2 = A_0^T = P_2 A_0 P_2^{-1}$ -- A_0^T and A_0 are similar.

More generally, assume A is similar to A_0. Then $A = PA_0P^{-1}$ for some nonsingular matrix P, and $x' = Ax$ is equivalent to $\psi(z)y = 0$ where $x = P\bar{x}$; $\psi(z)y = 0$ is equivalent to $\bar{x}' = A_0\bar{x}$, and this system is equivalent to $x' = Ax$ under the change of coordinates $x = P\bar{x}$. What we see is that every $\psi(z)y = 0$ is equivalent to a system $\bar{x}' = A_0\bar{x}$, and a system $x' = Ax$ *is equivalent to* an mth-order equation $\psi(z)y = 0$ *if and only if* A *is a companion matrix of* $\psi(z)$. As may already be evident, not every matrix is a companion matrix (the identity matrix, for example) and not every m-dimensional system $x' = Ax$ is equivalent to an mth-order equation. There is something special about companion matrices. For instance, the only ones for $m = 3$ have as their Jordan canonical form

$$\begin{pmatrix} \lambda_1 & 0 & 0 \\ 0 & \lambda_2 & 0 \\ 0 & 0 & \lambda_3 \end{pmatrix}, \quad (\lambda_3-\lambda_2)(\lambda_3-\lambda_1)(\lambda_2-\lambda_1) \neq 0;$$

$$\begin{pmatrix} \lambda_1 & 1 & 0 \\ 0 & \lambda_1 & 0 \\ 0 & 0 & \lambda_3 \end{pmatrix}, \quad \lambda \neq \lambda_3; \text{ or } \begin{pmatrix} \lambda_1 & 1 & 0 \\ 0 & \lambda_1 & 1 \\ 0 & 0 & \lambda_1 \end{pmatrix}.$$

Note in each case the geometric multiplicity of each eigenvalue is 1. The general proof of this for companion matrices is, we shall see, easy.

We now observe an interesting characterization of companion matrices. Suppose there is a vector b such that b, Ab, A^2b are linearly independent. Select b, Ab, A^2b as a basis for new coordinates; i.e., $x = P\hat{x}$, $P = (b, Ab, A^2b)$. The matrix $\hat{A} = P^{-1}AP$ represents in these coordinates the same linear transformation as A does in the original coordinates. Then

$$\hat{A} = \begin{pmatrix} 0 & 0 & -a_0 \\ 1 & 0 & -a_1 \\ 0 & 1 & -a_2 \end{pmatrix} = A_0^T,$$

where $\psi(\lambda) = \det(\lambda I - A) = \lambda^3 + a_2\lambda^2 + a_1\lambda + a_0$; $\hat{A}\begin{pmatrix} 1 \\ 0 \\ 0 \end{pmatrix}$ corresponds to Ab, $\hat{A}\begin{pmatrix} 0 \\ 1 \\ 0 \end{pmatrix}$ corresponds to A^2b and $\hat{A}\begin{pmatrix} 0 \\ 0 \\ 1 \end{pmatrix}$

corresponds to $A^3b = -a_0b - a_1Ab - a_2A^2b$. We say above that A_0^T is similar to A_0, and hence <u>the existence of a</u> b <u>such that</u> b, Ab, A^2b <u>are linearly independent</u> is a <u>sufficient condition for</u> A <u>to be a companion matrix</u>. Let us see that this condition is also necessary. Suppose A is similar to A_0. Then A is similar to A_1 under some change of coordinates $x = P\hat{x}$. Now

$$A_1^2 = \begin{pmatrix} \lambda_1^2 & \lambda_1+\lambda_2 & 1 \\ 0 & \lambda_2^2 & \lambda_2+\lambda_3 \\ 0 & 0 & \lambda_3^2 \end{pmatrix}.$$

Hence

$$\begin{pmatrix} 0 \\ 0 \\ 1 \end{pmatrix}, \quad A_1\begin{pmatrix} 0 \\ 0 \\ 1 \end{pmatrix} = \begin{pmatrix} 0 \\ 1 \\ \lambda_3 \end{pmatrix}, \quad A_1^2\begin{pmatrix} 0 \\ 0 \\ 1 \end{pmatrix} = \begin{pmatrix} 1 \\ \lambda_2+\lambda_3 \\ \lambda_3^2 \end{pmatrix},$$

and these three vectors are linearly independent. Therefore b, Ab, A^2b are linearly independent where b is the third column of P. The generalization of this result to dimension m plays, as we shall see in Section 15, an important role in control theory and is related to the controllability of linear systems.

Before turning to the general case let us introduce some notation and recall some basic concepts from linear algebra. Let $B = (b^1,\ldots,b^m)$ be any $m \times m$ (real or complex) matrix. The <u>kernel</u> (or <u>null space</u>) of B is $\ker B = \{x \in \mathbf{C}^m;\ Bx = 0\}$. The <u>image</u> (or <u>range</u>) of B is image $B = \{Bx;\ x \in \mathbf{C}^m\} = B(\mathbf{C}^m)$ and is the linear subspace spanned by $b^1,\ldots,b^m$. The dimension of the image of B is called the <u>rank</u> of B. A basic result in the theory of finite dimensional vector spaces is that $\dim(\ker B) + \text{rank } B = m$. For

$\lambda \in \sigma(A)$ let $\alpha(\lambda)$ denote the algebraic multiplicity of λ -- i.e., its multiplicity as a root of the characteristic polynomial of A. The number $\gamma(\lambda)$ of linearly independent eigenvectors of A associated with λ is called the _geometric multiplicity of_ λ; hence, $\gamma(\lambda) = \dim(\ker(A-\lambda I))$. Note also that $\gamma(\lambda)$ is the number of λ-blocks Q_i in the Jordan canonical form of A (blocks with λ along the diagonal). If $\alpha(\lambda) = \gamma(\lambda)$, then each λ-block is one dimensional, so that A is _diagonalizable_ (_semisimple_) _if and only if the geometric multiplicity of each eigenvalue of_ A _is equal to its algebraic multiplicity_. There is also a third multiplicity that can be associated with each eigenvalue; namely, its multiplicity $\mu(\lambda)$ as a root of the minimal polynomial. Hence, from the Jordan canonical form for A, we see that _the characteristic polynomial of_ A _is the minimal polynomial if and only if_ (i) $\alpha(\lambda) = \mu(\lambda)$ _or_ (ii) $\gamma(\lambda) = 1$ _for each_ $\lambda \in \sigma(A)$. Now $\gamma(\lambda) = \dim\ker(A-\lambda I) = m - \mathrm{rank}(A-\lambda I)$, and hence

10.1. _Proposition_. The characteristic polynomial of A is the minimal polynomial of A if and only if $\mathrm{rank}(A-\lambda I) = m-1$ for each $\lambda \in (A)$.

Consider the mth-order linear difference equation

$$y(n+m) + a_{m-1}y(n+m-1) + \cdots + a_0 y(n) = 0.$$

Letting

$$\psi(\lambda) = \lambda^m + a_{m-1}\lambda^{m-1} + \cdots + a_0 = \prod_{j=1}^{m} (\lambda-\lambda_j),$$

we can write this equation in the form

$$\psi(z)y = y^{(m)} + a_{m-1}y^{(m-1)} + \cdots + a_0 y = 0. \tag{10.8}$$

With

$$\overline{x} = \begin{pmatrix} y \\ y' \\ \vdots \\ y^{(m-1)} \end{pmatrix},$$

we see that (10.8) is equivalent to the system

$$\overline{x}' = A_0\overline{x}, \tag{10.9}$$

where

$$A_0 = \begin{pmatrix} 0 & 1 & 0 & \cdots & 0 \\ \vdots & & \ddots & \ddots & \vdots \\ & & & \ddots & 0 \\ 0 & \cdots & & 0 & 1 \\ -a_0 & -a_1 & \cdots & & -a_{m-1} \end{pmatrix}. \tag{10.10}$$

The vector $\overline{x}(n)$ is the state of (10.8) at time $n+m-1$. If $y(n)$ is a solution of (10.8), then $\overline{x}(n)$ is a solution of (10.9). Conversely, if $\overline{x}(n)$ is a solution of (10.9), then $y(n) = \overline{x}_1(n)$ is a solution of (10.8).

The matrix A_0 is called the <u>principal companion</u> matrix of the polynomial $\psi(\lambda)$. Any matrix similar to A_0 is called a <u>companion matrix</u> of $\psi(\lambda)$. Thus, to say that A is a companion matrix means that A is similar to a matrix of the form A_0 (equation (10.10); i.e., to the principal companion matrix A_0 of some polynomial $\psi(\lambda)$. We will see in a moment (Proposition 10.2) that $\psi(\lambda)$ is then the characteristic and minimal polynomial of A.

Now just as with $m = 3$ we have with

$$\begin{aligned}
x_1 &= y \\
x_2 &= (z-\lambda_1)y = p_{21}y + y' \\
x_3 &= (z-\lambda_2)(z-\lambda_1)y = p_{31}y + p_{32}y' + y'' \\
&\vdots \\
x_m &= (z-\lambda_{m-1})\cdots(z-\lambda_1)y = p_{m1}y + \cdots + p_{m,m-1}y^{(m-2)} + y^{m-1},
\end{aligned}$$

that, if y is a solution of (10.8), then x satisfies

$$\begin{aligned}
x_1' &= \lambda_1 x_1 + x_2 \\
x_2' &= \lambda_2 x_2 + x_3 \\
&\vdots \\
x_{m-1}' &= \lambda_{m-1}x_{m-1} + x_m \\
x_{m-1}' &= \lambda_m x_m .
\end{aligned}$$

Thus with

$$x = P_1\bar{x}, \quad P_1 = \begin{pmatrix} 1 & 0 & \cdots & 0 \\ p_{21} & \ddots & \ddots & \vdots \\ \vdots & \ddots & \ddots & 0 \\ p_{m1} & \cdots & p_{m,m-1} & 1 \end{pmatrix},$$

equation (10.8) is equivalent to

$$x' = A_1 x, \quad A_1 = \begin{pmatrix} \lambda_1 & 1 & 0 & \cdots & 0 \\ 0 & \ddots & \ddots & \ddots & \vdots \\ \vdots & & \ddots & \ddots & 0 \\ \vdots & & & \ddots & 1 \\ 0 & \cdots & & 0 & \lambda_m \end{pmatrix} \tag{10.11}$$

Also, just as for $m = 3$, with

$$x = P_2\bar{x}, \quad P_2 = \begin{bmatrix} a_1 & a_2 & & a_{m-1} & 1 \\ a_2 & a_3 & \cdots \; a_{m-1} & 1 & 0 \\ \vdots & & & & \vdots \\ a_{m-1} & & & & \\ 1 & 0 & \cdots & & 0 \end{bmatrix}, \tag{10.12}$$

equation (10.8) is equivalent to

$$x' = A_2 x, \quad A_2 = \begin{bmatrix} 0 & \cdots & 0 & -a_0 \\ 1 & \ddots & \vdots & -a_1 \\ 0 & \ddots & \ddots & \\ \vdots & \ddots & 0 & \vdots \\ 0 & \cdots & 0 \;\; 1 & -a_{m-1} \end{bmatrix} = A_0^T. \tag{10.13}$$

Note that $P_2 A_0 P_2^{-1} = A_0^T$. Thus A_1 and A_0^T are also companion matrices of $\psi(\lambda)$. Note since $\det(\lambda I - A_1)$ is obviously $\psi(\lambda)$, we have that $\psi(\lambda)$ is the characteristic polynomial of A_0. Later we will see it is also the minimal polynomial of A_0.

Although it is fairly clear what is meant by saying (*) $x' = Ax$ is equivalent to (**) $\psi(z)y = 0$, let us be precise. The system (*) is said to be <u>equivalent</u> to an mth-order equation (**) if there is a nonsingular matrix P such that if $y(n)$ is a solution of (**) then

$$x(n) = P \begin{bmatrix} y(n) \\ y'(n) \\ \vdots \\ y^{(m-1)}(n) \end{bmatrix} = P\bar{x}(n)$$

is a solution of (*) and, conversely, if $x(n)$ is a solution of (*), the first component $y(n)$ of $\bar{x}(n) = P^{-1}x(n)$ is a solution of (*); in other words, there is a one-to-one linear correspondence between solutions. We know that

$\bar{x}' = A_0\bar{x}$ is equivalent to (**), and hence a necessary and sufficient condition for a system $x' = Ax$ to be equivalent to an mth order difference equation $\psi(z)y = 0$ is that A be a companion matrix. Of course, $\psi(\lambda)$ is the characteristic polynomial of A and A is a companion matrix of $\psi(\lambda)$. This result is of interest as soon as we know how to identify companion matrices, and our objective now is to obtain a number of different characterizations of companion matrices.

10.2. Proposition. Let A be a companion matrix of $\psi(\lambda)$. Then $\psi(\lambda)$ is the characteristic and minimal polynomial of A.

Proof: Let A_0 (equation (10.10)) be the principal companion matrix of $\psi(\lambda)$. Since A is similar to A_0 and A_0 is similar to A_1 (equation (10.11)), what we need to show is that $\psi(\lambda)$ is the characteristic and minimal polynomial of A_1. It is obviously the characteristic polynomial. Omitting the first column and the last row in $A_1 - \lambda I$ we see that $\operatorname{rank}(A_1 - \lambda I) \geq m-1$. If $\lambda \in \sigma(A)$, then $\operatorname{rank}(A_1 - \lambda I) < m$, and hence $\operatorname{rank}(A_1 - \lambda I) = m-1$ for all $\lambda \in \sigma(A_1)$. Our result then follows from Proposition 10.1. □

10.3. Proposition. The following are equivalent (A any $m \times m$ matrix)

(i) A is a companion matrix.

(ii) $\operatorname{rank}(A - \lambda I) = m-1$ for each $\lambda \in \sigma(A)$.

(iii) the geometric multiplicity of each eigenvalue of A is 1.

(iv) The characteristic polynomial of A is the minimal polynomial of A.

(v) There is a vector b such that $b, Ab, \ldots, A^{m-1}b$ are linearly independent.

Proof: Assume (iv), and let $\psi(\lambda)$ be the characteristic polynomial of A and A_0 its principal companion matrix. By Proposition 10.2, $\psi(\lambda)$ is the characteristic and minimal polynomial of A_0. It is then relatively easy to see that A and A_0 have the same Jordan canonical form and are therefore similar. Hence (iv) implies (i). From Proposition 10.1 and the discussion above it we know that (ii), (iii) and (iv) are equivalent. Proposition 10.2 tells us that (i) implies (iv), and (i) through (iv) are equivalent.

The proof of the equivalence of (i) and (v) is just like $m = 3$. Assume (v) and take $b, Ab, \ldots, A^{m-1}b$ as the basis of new coordinates -- $x = P\hat{x}$, $P = (b, Ab, \ldots, A^{m-1}b)$. Then $\hat{A} = P^{-1}AP = A_0^T$, which we know is similar to A_0; i.e., (v) implies (i). Assume (i). Then A is similar to A_1 (equation (10.11)). It is relatively easy to see that (look at the last columns of $A_1, A_1^2, \ldots, A_1^{m-1}$)

$$\delta^m = \begin{pmatrix} 0 \\ \vdots \\ 0 \\ 1 \end{pmatrix}, \qquad A_1\delta^m = \begin{pmatrix} 0 \\ \vdots \\ 0 \\ 1 \\ \lambda_m \end{pmatrix},$$

$$A_1^2\delta^m = \begin{pmatrix} 0 \\ \vdots \\ 0 \\ 1 \\ \lambda_m + \lambda_{m-1} \\ \lambda_m^2 \end{pmatrix}, \ldots, A^{m-1}\delta^m = \begin{pmatrix} 1 \\ --- \\ --- \\ \vdots \\ --- \\ \lambda_m^{m-1} \end{pmatrix}$$

where the blanks are functions of the λ_i;

$$(\delta^m, A_1\delta^m, \ldots, A_1^{m-1}\delta^m) = \begin{pmatrix} 0 & \cdots & 0 & 1 \\ \vdots & & \cdot^{\cdot^{\cdot}} & - \\ 0 & \cdot^{\cdot^{\cdot}} & & - \\ 1 & - & - & - \end{pmatrix}$$

and hence $\delta^m, A_1\delta^m, \ldots, A_1^{m-1}\delta^m$ are linearly independent. Therefore (i) implies (v) with $b = P^m$ -- the last column is P -- where $A = P^{-1}A_1P$. □

We have finally (from our observation above Proposition 10.2) our basic result (in conjunction with Proposition 10.3).

10.4. <u>Theorem.</u> A system $x' = Ax$ is similar to an mth-order equation $\psi(z)y = 0$ if and only if $\operatorname{rank}(A-\lambda I) = m-1$ for each $\lambda \in \sigma(A)$.

We also know that, if $x' = Ax$ is equivalent to $\psi(z)y = 0$, then $\psi(\lambda)$ is the characteristic and minimal polynomial of A. What we have done here applies equally well to systems of differential equations $\frac{dx}{dt} = Ax$ and m^{th}-order differential equations

$$\psi(D)y = \frac{d^m y}{dt^m} + a_{m-1}\frac{d^{m-1}y}{dt^{m-1}} + \cdots + a_0 y = 0 .$$

Let us look at one application of our results whose significance, if you are new to control theory, may not be immediately obvious.

We will now prove the following proposition (it is related to what engineers call in system theory "pole assignment").

10.5. <u>Proposition</u>. Let A be a real $m \times m$ matrix, $\sigma_0 = \{\lambda_1, \ldots, \lambda_m\}$ an arbitrary set of eigenvalues (complex numbers) with the property that $\overline{\sigma}_0 = \{\overline{\lambda}_1, \ldots, \overline{\lambda}_m\} = \sigma_0$, and let b be any vector in $\mathbf{R}^m$. If $b, Ab, \ldots, A^{m-1}b$ are linearly independent, then corresponding to each σ_0 there is a $c \in \mathbf{R}^m$ such that $\sigma(A+bc^T) = \sigma_0$.

<u>Proof</u>: Consider the system

$$x' = Ax + bu(x), \tag{10.14}$$

where $u: R^m \to R$. Since $b, Ab, \ldots, A^{m-1}b$ are linearly independent, we see from the proof of Proposition 10.3 that under the change of coordinates $x = P\hat{x}$, $P = (b, Ab, \ldots, A^{m-1}b)$, (10.14) becomes $\hat{x}' = A_0^T\hat{x} + P^{-1}bu(P\hat{x})$. But $P\delta^1 = b$, and

$$\hat{x}' = A_0^T\hat{x} + \delta^1 v(\hat{x}),$$

where $v(\hat{x}) = u(P\hat{x})$. With $\hat{x} = P_2\overline{x}$, P_2 given by equation (10.12), we see that

$$\overline{x}' = A_0\overline{x} + \delta^m w(\overline{x}),$$

since $P_2\delta^m = \delta^1$, and (9.14) is equivalent to

$$\psi(z)y = y^{(m)} + a_{m-1}y^{(m-1)} + \cdots + a_0 y = w(\overline{x}). \tag{10.15}$$

Taking $w(\overline{x}) = d^T\overline{x} = d_1 y + d_2 y' + \cdots + d_m y^{(m-1)}$, we see that

$$x' = Ax + bc^T x$$

is equivalent to

$$\psi_0(z)y = y^{(m)} + (a_{m-1} - d_m)y^{(m-1)} + \cdots + (a_0 - d_1)y = 0,$$

where $d = P_2^T P^T c$. Since the characteristic polynomial of

$A+bc^T$ is $\psi_0(\lambda)$, the conclusion of the proposition is obvious. □

10.5. Exercise. Prove the converse of Proposition 10.5.

This proposition means, for instance, if $x' = Ax$ is an uncontrolled system, and if only one component of control can be applied (Figure 10.1), then the system $x' = Ax + bu$

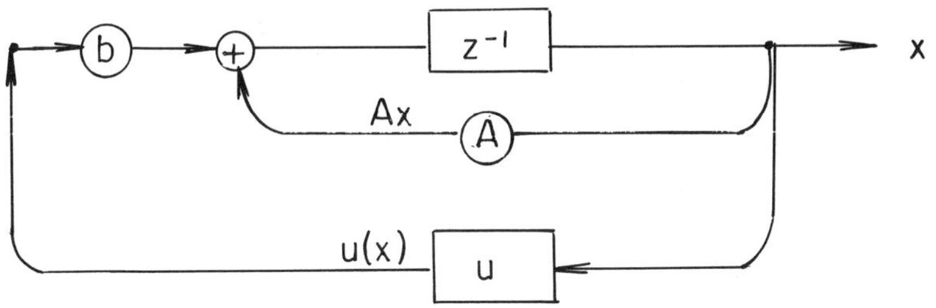

Feedback Control

Figure 10.1

can be stabilized by linear feedback control $u(x) = c_1x_1 + c_2x_2 + \cdots + c_nx_n$ if $b, Ab, \ldots, A^{m-1}b$ are linearly independent. We will see in Section 15 that this is the condition for complete controllability of the system (10.14). Not only can the system be stabilized the eigenvalues of $A+bc$ can be arbitrarily assigned. For instance, one could choose c so that all of the eigenvalues of $A+bc$ are zero. Then $(A+bc)^m = 0$, and all solutions of the controlled system reach the origin in time m.

11. Another algorithm for computing A^n.

In Section 4 we gave an algorithm for computing A^n that depended upon computing the eigenvalues of A. Here in this section we give an algorithm that does not require computing the eigenvalues. As before we let

$$\psi(\lambda) = \lambda^s + a_{s-1}\lambda^{s-1} + \cdots + a_0$$

be any polynomial that annihilates A -- i.e., such that $\psi(A) = 0$. We can, for instance, always take $\psi(\lambda)$ to be the characteristic polynomial of A.

If $x(n)$ is a solution of

$$x' = Ax, \tag{11.2}$$

then $\psi(z)x(n) = \psi(A)x(n) = 0$, and $x(n)$ is a solution of

$$\psi(z)x = 0. \tag{11.3}$$

This vector s^{th}-order equation we can solve, and we know that all of the solutions of (11.2) are among those of (11.3). One can proceed in this manner and obtain an algorithm, but it is simpler to do much as we did in Section 4.

Since

$$A^s = -a_0I - a_1A - \cdots - a_{s-1}A^{s-1},$$

we see that

$$A^n = \sum_{j=1}^{s} u_j(n)A^{j-1}. \tag{11.4}$$

If $\psi(\lambda)$ were the minimal polynomial of A, then $I, A, \ldots, A^{s-1}$ would be linearly independent, and the $u_j(n)$ would be uniquely determined by (10.4). However, all that is necessary is to select $u_j(n)$ so that $X(0) = I$ and $AX(n) = X(n+1)$, where $X(n) = \sum_{j=1}^{s} u_j(n)A^{j-1}$. Examining these equations we obtain immediately that these conditions are satisfied if

$$u_1(n+1) = -a_0 u_s(n) \tag{11.5}$$

$$u_j(n+1) = u_{j-1}(n) - a_{j-1}u_s(n), \; j = 2,\dots,s$$

and

$$u_1(0) = 1, \; u_j(0) = 0, \quad j \geq 2. \tag{11.6}$$

Thus $X(n) = A^n$ and to compute A^n one can take $\psi(\lambda)$ to be the characteristic polynomial of A (which makes $s = m$), compute the coefficients $a_0,\dots,a_{m-1}$, the products $A, A^2,\dots,A^{m-1}$, and starting with the initial conditions (11.6) use (11.5) to compute the $u_j(n)$.

Writing

$$u = \begin{pmatrix} u_1 \\ \vdots \\ u_s \end{pmatrix},$$

equation (11.5) becomes

$$u' = Bu, \tag{11.7}$$

where

$$B = \begin{pmatrix} 0 & & & 0 & -a_0 \\ 1 & \ddots & & & -a_1 \\ 0 & \ddots & \ddots & & \vdots \\ \vdots & & \ddots & 0 & \vdots \\ 0 & \dots & 0 & 1 & -a_{s-1} \end{pmatrix}$$

Hence the $u(n)$ of equation (11.4) are the components of the first principal solution of (11.7) -- $u(n)$ is the first column of B^n. Another way to look at this is that the computation of A^n has been reduced to computing B^n. Since

B is a companion matrix of $\psi(\lambda)$, (11.7) is equivalent to the sth order equation $\psi(z)y = 0$; i.e.

$$y^{(s)} + a_{s-1}y^{(s-1)} + \cdots + a_0 y = 0. \tag{11.8}$$

In fact we know from Section 10 that the solutions of (11.7) and (11.8) are related by

$$\begin{aligned} u_1 &= a_1 y + a_2 y' + \cdots + a_{s-1}y^{(s-2)} + y^{(s-1)} \\ u_2 &= a_2 y + a_3 y' + \cdots + y^{(s-2)} \\ &\vdots \\ u_{s-1} &= a_{s-1}y + y' \\ u_s &= y \end{aligned} \tag{11.9}$$

or

$$u = \begin{pmatrix} a_1 & a_2 & \cdots & a_{s-1} & 1 \\ a_2 & & & \cdot & 0 \\ \vdots & & \cdot & & \vdots \\ a_{s-1} & \cdot & & & \\ 1 & 0 & \cdots & & 0 \end{pmatrix} \begin{pmatrix} y \\ y' \\ \vdots \\ y^{(s-1)} \end{pmatrix} \tag{11.10}$$

Note that the initial condition

$$\begin{pmatrix} y(0) \\ y'(0) \\ \vdots \\ y^{(s-1)}(0) \end{pmatrix} = \begin{pmatrix} 0 \\ 0 \\ \vdots \\ 0 \\ 1 \end{pmatrix} \quad \text{corresponds to} \quad u(0) = \begin{pmatrix} 1 \\ 0 \\ \vdots \\ 0 \end{pmatrix},$$

so that the u_j's can be computed by computing the last principal solution of (11.8).

11.1. Example. Find the principal solutions of

$$x_1' = \frac{3}{4}x_1 - \frac{1}{4}x_2$$

$$x_2' = \frac{1}{4}x_1 + \frac{1}{4}x_2$$

$$x_3' = \frac{1}{4}x_1 - \frac{1}{4}x_2 + \frac{1}{2}y_3.$$

<u>Solution 1</u>. (By the algorithm above).

$$A = \frac{1}{4}\begin{pmatrix} 3 & -1 & 0 \\ 1 & 1 & 0 \\ 1 & -1 & 2 \end{pmatrix}$$

Take $\psi(\lambda) = \det(\lambda I - A) = (\lambda - \frac{1}{2})^3 = \lambda^3 - \frac{3}{2}\lambda^2 + \frac{3}{4}\lambda - \frac{1}{8}$. The 3rd principal solution of

$$y''' - \frac{3}{2}y'' + \frac{3}{4}y' - \frac{1}{8}y = 0$$

is

$$y(n) = n(n-1)2^{-n+1}.$$

Now

$$u_1(n) = \frac{3}{4}y^{(n)} - \frac{3}{2}y'(n) + y''(n) = (n-1)(n-2)2^{-n-1}$$

$$u_2(n) = -\frac{3}{2}y(n) + y'(n) = -n(n-2)2^{-n+1}$$

$$u_3(n) = y(n) = n(n-1)2^{-n+1},$$

and

$$A^2 = \frac{1}{4}\begin{pmatrix} 2 & -1 & 0 \\ 1 & 0 & 0 \\ 1 & -1 & 1 \end{pmatrix}$$

Then

$$A^n = u_1(n)I + u_2(n)A + u_3(n)A^2$$

$$= \frac{1}{2^{n+1}}\begin{pmatrix} 2+n & -n & 0 \\ n & 2-n & 0 \\ n & -n & 2 \end{pmatrix}.$$

Therefore the principal solutions are

$$x^1(n) = \frac{1}{2^{n+1}}\begin{vmatrix} 2+n \\ n \\ n \end{vmatrix}, \quad x^2(n) = \frac{1}{2^{n+1}}\begin{vmatrix} -n \\ 2-n \\ -n \end{vmatrix}$$

$$\text{and, } x^3(n) = \frac{1}{2^n}\begin{vmatrix} 0 \\ 0 \\ 1 \end{vmatrix}.$$

Note, however, that here we used our knowledge of the eigenvalues to obtain $y(n)$.

Solution 2. (By the algorithm of Section 4).

Here $Q_1 = A - \frac{1}{2}I = \frac{1}{4}\begin{vmatrix} 1 & -1 & 0 \\ 1 & -1 & 0 \\ 1 & -1 & 0 \end{vmatrix}$ and $Q_2 = (A - \frac{1}{2}I)^2 = 0$.

From equations (4.4) $w_1(n) = 2^{-n}$ and $w_2(n) = n2^{-n+1}$. Then

$$A^n = w_1(n)I + w_2(n)Q_1 = \frac{1}{2^{n+1}}\begin{vmatrix} 2+n & -n & 0 \\ n & 2-n & 0 \\ n & -n & n \end{vmatrix}.$$

When it is easy to compute the eigenvalues the first algorithm (Section 4) will be simpler to use.

11.2. Exercise. Use both algorithms to derive a formula for A^n where

$$A = \begin{vmatrix} 1 & 0 & -1 \\ 0 & 1 & -2 \\ -1 & 1 & 1 \end{vmatrix}.$$

12. Nonhomogeneous linear systems $x' = Ax + f(n)$. Variation of parameters and undetermined coefficients.

The general nonhomogeneous linear system with constant coefficients is

$$x' = Ax + f(n) \tag{12.1}$$

where, as always in this chapter, A is an $m \times m$ real matrix and $f: J_0 \to \mathbf{C}^m$. If $f(n) = f_1(n) + if_2(n)$, where $f_1(n)$ and $f_2(n)$ are real, and if $x(n) = x^1(n) + ix^2(n)$ is a solution of (12.1), $x^1(n)$ and $x^2(n)$ real, then $x^{1'}(n) = Ax^1(n) + f_1(n)$ and $x^{2'}(n) = Ax^2(n) + f_2(n)$; and conversely, if $x^1(n)$ and $x^2(n)$ are real solutions of $x^{1'} = Ax^1 + f_1(n)$ and $x^{2'} = Ax^2 + f_2(n)$, then $x(n) = x^1(n) + ix^2(n)$ is a solution of (12.1). Thus, it is no more general to consider complex valued $f(n)$, but it is convenient to do so. The block diagram for (12.1) is shown in Figure 12.1.

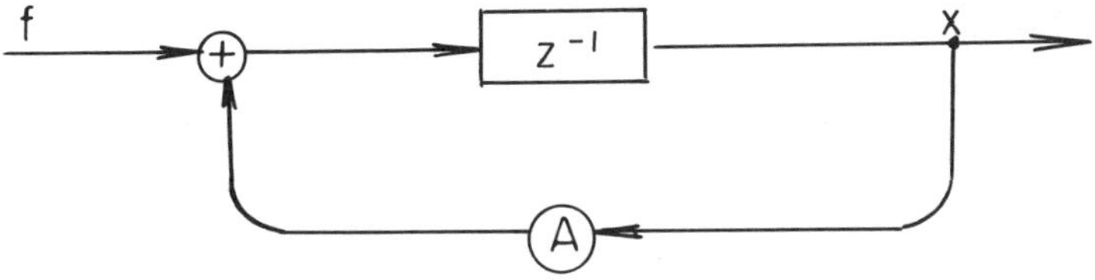

Figure 12.1

The function f is the <u>input</u>, or forcing term, and x is the <u>output</u>. The <u>superposition principle</u> is that, if $x^1(n)$ and $x^2(n)$ satisfy

$$x^{1'} = Ax^1 + f_1(n)$$

and

$$x^{2'} = Ax^2 + f_2(n),$$

then $x(n) = c_1x^1(n) + c_2x^2(n)$ is a solution of

$$x' = Ax + c_1f_1(n) + c_2f_2(n).$$

For instance, if the mapping $\mathscr{L}$ of the space of functions $f: J_0 \to \mathbf{C}$ into itself is defined by $\mathscr{L}(f) = x$, where x is the solution of (12.1) satisfying $x(0) = 0$, then $\mathcal{L}$ is linear; i.e. $\mathcal{L}(c_1 f_1 + c_2 f_2) = c_1 \mathcal{L}(f_1) + c_2 \mathcal{L}(f_2)$.

From the superposition principle we have, if $x^1(n)$ and $x^2(n)$ are solutions of (12.1), then $x^1(n) - x^2(n)$ is a solution of the homogeneous equation

$$x' = Ax. \tag{12.2}$$

Thus, if $x^1(n)$ is a particular solution of the complete equation (12.1), the general solution of (12.1) is $A^n c + x^1(n)$ -- <u>the general solution of the complete equation is the general solution of the homogeneous equation plus any particular solution of the complete equation.</u>

12.1. <u>Example</u>. Let us solve

$$4x'' + 4x' + x = \cos \frac{n\pi}{2} . \tag{12.3}$$

The input or forcing term $\cos \frac{n\pi}{2}$ is periodic of period 4 (the sequence of values of $\cos \frac{n\pi}{2}$ is $1,0,-1,0,1,\ldots$). Now $e^{i\frac{n\pi}{2}} = \cos \frac{n\pi}{2} + i \sin \frac{n\pi}{2}$ so that, if we solve

$$4x'' + 4x' + x = e^{i\frac{n\pi}{2}}, \tag{12.4}$$

the real part of this solution is a solution of (12.3). We look for a particular solution of (12.4) of the form $\hat{p}(n) = ce^{i\frac{n\pi}{2}}$. Substituting into (12.4) gives

$$(4e^{i\pi} + 4e^{i\frac{\pi}{2}} + 1)c = (-3+4i)c = 1 \quad \text{or } c = -\frac{3+4i}{25} .$$

Hence a particular solution of (12.3) is

$$p(n) = \operatorname{Re}\left(-\frac{3+4i}{25}e^{i\frac{n\pi}{2}}\right) = -\frac{3}{25}\cos \frac{n\pi}{2} + \frac{4}{25}\sin \frac{n\pi}{2} .$$

The characteristic polynomial of (12.3) is $\phi(\lambda) = 4\lambda^2 + 4\lambda + 1 = (2\lambda+1)^2$, and the general solution of (12.3) is

$$x(n) = (-1)^n(c_1+nc_2)2^{-n} - \frac{3}{25}\cos\frac{n\pi}{2} + \frac{4}{25}\sin\frac{n\pi}{2} .$$

12.2. <u>Exercise</u>. Solve $4x'' + 4x' + x = \sin\frac{n\pi}{4}$.

We now rederive the so-called variation of constants formula for (12.1). Let $x(n)$ be the solution of (12.1) satisfying $x(0) = x^0$. Then A^{n-k}, $n \geq k$, is an integrating factor of $x'(k) - Ax(k) = f(k)$; i.e.

$$A^{n+1-(k+1)}x(k+1) - A^{n+1-k}x(k) = A^{n-k}f(k),$$

and summing from $k = 0$ to $k = n-1$ we obtain

$$x(n) = A^n x(0) + \sum_{k=0}^{n-1} A^{n-k-1}f(k). \qquad (12.5)$$

12.3. <u>Proposition</u> (<u>Variation of Constants Formula</u>). The solution of (12.1) satisfying $x(0) = x^0$ is given by (12.5).

It is useful to note that $\sum_{k=0}^{n-1} A^{n-k-1}f(k) = \sum_{k=0}^{n-1} A^k f(n-k-1)$, and the variation of constants formula can also be written

$$x(n) = A^n x(0) + \sum_{k=0}^{n-1} A^k f(n-k-1). \qquad (12.6)$$

Clearly, $\sum_{k=0}^{n-1} A^{n-k-1}f(k)$ is the solution of (12.1) satisfying $x(0) = 0$. If the system is started at time $n = 0$ from rest at the origin, then $\sum_{k=0}^{n-1} A^{n-k-1}f(k)$ is its output.

The particular case of the variation of constants formula for the mth-order equation

$$\psi(z)y = y^{(m)} + a_{m-1}y^{(m-1)} + \cdots + a_0 y = u(n) \qquad (12.7)$$

is given in the next exercise.

12.4. Exercise. The solution $y(n)$ of (12.7) satisfying $y(0) = y(1) = \cdots = y(m-1) = 0$ is given by

$$y(n) = \sum_{k=0}^{n-1} w(n-k-1)u(k) = \sum_{k=0}^{n-1} w(k)u(n-k-1), \qquad (12.8)$$

where $w(n)$ is the mth-principal solution of $\psi(z)y = 0$; i.e., the solution satisfying $y(0) = y(1) = \cdots = y(m-2) = 0$, $y(m-1) = 1$.

12.5. Example. Let us find the solution of $x'' - \sqrt{2}\,x' + x = \log(n+1)$ satisfying $x(0) = x(1) = 0$. The characteristic polynomial is

$$\phi(\lambda) = \lambda^2 - \sqrt{2}\,\lambda + 1 = \left(\lambda - \frac{1+i}{\sqrt{2}}\right)\left(\lambda - \frac{1-i}{\sqrt{2}}\right)$$

so that the general solution of

$$x'' - \sqrt{2}\,x' + x = 0$$

is

$$x(n) = c_1\left(\frac{1+i}{\sqrt{2}}\right)^n + c_2\left(\frac{1-i}{\sqrt{2}}\right)^n$$

$$= c_1 e^{i\frac{n\pi}{4}} + c_2 e^{-i\frac{n\pi}{4}} .$$

The solution satisfying $x(0) = 0$, $x(1) = 1$, is given by $c_1 = -\frac{i}{2}$ and $c_2 = \frac{i}{2}$, i.e. the 2nd-principal solution is

$$w(n) = -\frac{i}{2}e^{i\frac{n\pi}{4}} + \frac{i}{2}\,e^{-i\frac{n\pi}{4}} = \sqrt{2}\ \sin\frac{n\pi}{4} .$$

By (12.8) the solution we seek is

$$x(n) = \sqrt{2}\sum_{k=0}^{n-2} \sin(n-k-1)\frac{\pi}{4}\,\log(k+1)$$

$$= \sqrt{2}\sum_{k=1}^{n-1} \sin\frac{k\pi}{4}\,\log(n-k).$$

12.6. Exercise. (see Exercise 12.1). Show that

$$\sum_{k=0}^{n-2} (-1)^{n-k}(n-k-1)2^{k-n} \cos \frac{k\pi}{2}$$

$$= \sum_{k=1}^{n-1} (-1)^{k+1}k2^{-k+1} \cos(n-k-1)\frac{\pi}{2}$$

$$= (-1)^n \frac{1}{25}(3+5n)2^{-n} - \frac{3}{25} \cos \frac{n\pi}{2} + \frac{4}{25} \sin \frac{n\pi}{2} .$$

We now want to look at what is called the method of undetermined coefficients for obtaining a particular solution of

$$x' = Ax + p(n)\lambda^n \tag{12.9}$$

where $\lambda \in \mathbf{C}$ and

$$p(n) = \sum_{j=0}^{r} n^j a^j = \sum_{j=0}^{r} \binom{n}{j} b^j ;$$

$a^j, b^j \in \mathbf{C}^m$ and $p(n)$ is a vector polynomial of degree r. There certainly must be a polynomial solution $x(n) = q(n)\lambda^n$ where $q(n) = \sum_{j=0}^{r_0} \binom{n}{j} c^j$ is a vector polynomial. The method of undetermined coefficients for finding a particular solution of (12.9) is then to substitute $q(n)$ into (12.9) and determine the coefficients c^j. Consider, for instance,

$$x' = Ax + \lambda^n b, \tag{12.10}$$

and look for a solution of the form $x(n) = \lambda^n c$. Substituting into (12.10) gives

$$(\lambda I-A)c = b,$$

and we see immediately that, if λ is not an eigenvalue of $A(\lambda \notin \sigma(A))$, $c = (\lambda I-A)^{-1}b$.

12.7. Example. Let us find the general solution of

$$x'' - 2(\cos \omega_0)x' + x = a \cos n\omega. \tag{12.11}$$

We replace this equation by

$$x'' = 2(\cos \omega_0)x' - x + ae^{in\omega}$$

and look for a solution of the form $x(n) = ce^{in\omega}$. Substituting gives

$$\phi(e^{i\omega})c = a$$

where

$$\phi(\lambda) = \lambda^2 - 2(\cos \omega_0)\lambda + 1 = (\lambda - e^{i\omega_0})(\lambda - e^{-i\omega_0})$$

is the characteristic polynomial. Hence, if $e^{i\omega}$ is not an eigenvalue, then $c = a/\phi(e^{i\omega})$. We may assume $0 \le \omega \le \pi$ and $0 \le \omega_0 \le \pi$, andobtain as a particular solution when $\omega \neq \omega_0$

$$x(n) = \mathrm{Re}\left(\frac{ae^{in\omega}}{\phi(e^{i\omega})}\right) = \frac{a}{\rho(\omega)} \cos(n\omega - \delta(\omega)),$$

where $\phi(e^{i\omega}) = \rho(\omega)e^{i\delta(\omega)}$. The general solution of (12.11) is therefore

$$x(n) = c_1 \cos \omega_0 n + c_2 \sin \omega_0 n + \frac{a}{\rho(\omega)} \cos(n\omega - \delta(\omega)), \ \omega \neq \omega_0.$$

When $\omega = \omega_0$, the forcing term is a solution of the homogeneous equation and this is like resonance. As $\omega \to \omega_0$, $\frac{a}{\rho(\omega)} \to \infty$. We will consider this case in a moment (Example 12.10). Note that the domain of definition of our functions is J_0, the nonnegative integers, and $e^{in\omega}$ is periodic if and only if ω is a rational multiple of 2π.

In general we must consider two cases -- $\lambda \notin \sigma(A)$ and $\lambda \in \sigma(A)$. Let us look at $\lambda \notin \sigma(A)$ first. By the superposition principle we need consider only

$$x' = Ax + \binom{n}{r}\lambda^n b, \tag{12.12}$$

and we look for a solution of the form

$$x(n) = \lambda^n \sum_{j=0}^{r} \binom{n}{j} c^j .$$

Substituting into the equation gives $(\lambda \neq 0)$

$$\lambda \sum_{j=0}^{r} \left[\binom{n+1}{j} - \binom{n}{j}\right] c^j = \sum_{j=0}^{r} \binom{n}{j}(A-\lambda I)c^j + \binom{n}{r} b$$

or

$$\sum_{j=0}^{r} \left[\binom{n}{j}(\lambda I - A) + \lambda\binom{n}{j-1}\right] c^j = \binom{n}{r} b .$$

Hence

$$(\lambda I - A)c^r = b$$
$$(\lambda I - A)c^{j-1} + \lambda c^j = 0, \quad j = r,\ldots,1,$$

and

$$c^{r-j} = \lambda^j (A-\lambda I)^{-j-1} b .$$

What we see is that

$$x(n) = \lambda^n \sum_{j=0}^{r} \binom{n}{r-j} \lambda^j (A-\lambda I)^{-j-1} b$$

is a solution of (12.12) when λ is not an eigenvalue of A (without the restriction $\lambda = 0$).

The case $\lambda \in \sigma(A)$ is a bit more complicated. Let us assume that we have written equation (12.10) in Jordan canonical form and that the blocks have been so ordered that $\lambda = \lambda_i$, $i = 1,\ldots,j$, and $\lambda \neq \lambda_i$, $i = j+1,\ldots,s$. Let

$$A_1 = \operatorname{diag}\{Q_1,\ldots,Q_j\}$$

and

$$A_2 = \operatorname{diag}\{Q_{j+1},\ldots,Q_s\}.$$

Then with $x = \binom{u}{w}$ our system has been decoupled, and our problem is reduced to considering

$$u' = A_1 u + \binom{n}{r}\lambda^n b^1$$

$$v' = A_2 v + \binom{n}{r}\lambda^n b^2,$$

where $A_2 - \lambda I$ is nonsingular and $A_1 = \lambda I + N$; $N^k = 0$, $k = \max\{s_1, \ldots, s_j\}$. We already know that the second of these equations has a solution that is a polynomial of degree r times λ^n. Going through the same procedure as before, it is easy to see that

$$u(n) = \sum_{j=0}^{k-1} \binom{n}{r+j+1}\lambda^{n-j-1} N^j b^1$$

is a solution of the first equation. We may not know k, but we do know that $k \leq \alpha(\lambda)$, the algebraic multiplicity of λ. This particular solution is λ^n times a polynomial of degree $r+k$. Summing up, we have shown that

12.8. <u>Proposition</u>. The equation $x' = Ax + p(n)\lambda^n$, where p is a vector polynomial of degree r has a solution of the form $x(n) = q(n)\lambda^n$ where q is vector polynomial of degree r_0. If λ is not an eigenvalue of A, $r_0 = r$. If λ is an eigenvalue of A, then r_0 is less than or equal to r plus the algebraic multiplicity of λ.

In the particular case of an mth order equation

$$y^{(m)} + a_{m-1}y^{(m-1)} + \cdots + a_0 y = p(n)\lambda^n, \tag{12.13}$$

where $p(n)$ is a polynomial of degree r, we know that the k above is the algebraic multiplicity of λ. We also know for for any polynomial p_0 of degree less than k that p_0 $p_0(n)\lambda^n$ is a solution of the homogeneous equation. Hence

12.9. <u>Proposition</u>. Equation (12.13) has a solution of the form $n^\alpha q(n)\lambda^n$ where α is the algebraic multiplicity of λ

and $q(n)$ is a polynomial of degree r.

12.10. <u>Example</u>. Let us consider the resonant like case $\omega = \omega_0$ for (12.11) -- $x'' - 2(\cos \omega_0)x' + x + a \cos n\,\omega_0$. We know that $x'' - 2\cos \omega_0 x + x = ae^{in\omega_0}$ has a solution of the form $x(n) = cne^{in\omega_0}$. Substituting into the equation gives

$$c^{i\omega_0}(e^{i\omega_0} - \cos \omega_0)c = a \quad \text{and} \quad c = \frac{ce^{-i\omega_0}}{2i \sin \omega_0} a.$$

Hence

$$x(n) = R\ell\left(\frac{e^{i(n-1)\omega_0}}{2i \sin \omega_0} a\right) = na\,\frac{\sin(n+1)\omega_0}{2 \sin \omega_0}$$

is a solution of the original equation and the general solution is

$$x(n) = c_1 \cos \omega_0 n + c_2 \sin \omega_0 n + na\,\frac{\sin(n+1)\omega_0}{2 \sin \omega_0}.$$

For $\omega_0 = 0$, the general solution of

$$x''-2x'+x = a \quad \text{is} \quad x(n) = c_1+c_2 n+ \frac{a}{2}\,n(n+1).$$

Thus for $a \neq 0$ all solutions are unbounded when $\omega = \omega_0$.

12.11. <u>Example.</u> We now use the method of undetermined coefficients to solve

$$\begin{aligned} 4x_1' &= x_1 - x_2 - x_3 + (n+1)2^{-2n} \\ 4x_2' &= x_1 + x_2 + 2^{-n} \\ 4x_3' &= -x_1 + x_2 + 2x_3 + n2^{-n} \end{aligned} \tag{12.14}$$

Hence

$$4x' = Ax + f(n),$$

where

$$A = \begin{pmatrix} 1 & -1 & -1 \\ 1 & 1 & 0 \\ -1 & 1 & 2 \end{pmatrix}$$

and

$$f(n) = 2^{-n}\begin{pmatrix} 0 \\ 1 \\ n \end{pmatrix} + 2^{-2n}\begin{pmatrix} n+1 \\ 0 \\ 0 \end{pmatrix}.$$

$$\phi(\lambda) = \det(\lambda I - \frac{1}{4}A) = (\lambda - \frac{1}{2})(\lambda - \frac{1}{4})^2$$

and the eigenvalues of our problem are $\frac{1}{2}, \frac{1}{4}, \frac{1}{4}$ -- the eigenvalues of A are $2, 1, 1$.

We solve first

$$4x' = Ax + 2^{-n}p(n), \quad p(n) = \begin{pmatrix} 0 \\ 1 \\ n \end{pmatrix}.$$

We look for -- and know we will find -- a solution of the form

$$x(n) = q^{-n}a(n), \quad q(n) = c^0 + nc^1 + \frac{1}{2}n(n-1)c^2.$$

Substituting into the equation, we have $(\Delta q = q'-q)$

$$\frac{1}{2}\,4\Delta q = 2\Delta q = (A-2I)q + p$$

or

$$c^1 + nc^2 = \sum_{j=0}^{2} \binom{n}{j}(A-2I)c^j + \begin{pmatrix} 0 \\ 1 \\ n \end{pmatrix}$$

The equations to be solved are

$$(A-2I)c^2 = 0$$

$$(A-2I)c^1 = c^2 - \begin{pmatrix} 0 \\ 0 \\ 1 \end{pmatrix}$$

$$(A-2I)c^0 = c^1 - \begin{pmatrix} 0 \\ 1 \\ 0 \end{pmatrix};$$

$$(A-2I) = \begin{pmatrix} -1 & -1 & -1 \\ 1 & -1 & 0 \\ -1 & 1 & 0 \end{pmatrix}.$$

Solving these equations in succession gives

$$c^2 = -\frac{1}{2}\begin{pmatrix} 1 \\ 1 \\ -2 \end{pmatrix}, \quad c^1 = \frac{1}{2}\begin{pmatrix} 1 \\ 3 \\ -1 \end{pmatrix}, \quad c^0 = \begin{pmatrix} -1 \\ -3 \\ 3 \end{pmatrix},$$

and a particular solution is

$$x(n) = 2^{-n}\left[\begin{pmatrix} -1 \\ -3 \\ 3 \end{pmatrix} + \frac{1}{2}n\begin{pmatrix} 1 \\ 3 \\ -1 \end{pmatrix} - \frac{1}{4}\,n(n-1)\begin{pmatrix} 1 \\ 1 \\ -2 \end{pmatrix}\right].$$

The vector $v^1 = \begin{pmatrix} 1 \\ 1 \\ -2 \end{pmatrix}$ is an eigenvector of A corresponding to the eigenvalue 2.

We next solve

$$4x' = Ax + 2^{-2n}p(n), \quad p(n) = \begin{pmatrix} n+1 \\ 0 \\ 0 \end{pmatrix}. \tag{12.15}$$

Since $\frac{1}{4}$ is a double eigenvalue $\frac{1}{4}A$, we take

$$q(n) = \sum_{j=0}^{3} \binom{n}{j} c^j.$$

We want to find v^2 and v^3 satisfying $(A-I)v^2 = 0$ and $(A-I)^2 v^3 = 0$; solutions are

$$v^2 = \begin{pmatrix} 0 \\ 1 \\ -1 \end{pmatrix}, \quad v^3 = \begin{pmatrix} 1 \\ 0 \\ 0 \end{pmatrix}.$$

Thus, we might as well take v^1, v^2, and v^3 as a basis of new coordinates and reduce A to Jordan canonical form; i.e., $x = P\hat{x}$, where

$$P = \begin{pmatrix} 1 & 0 & 1 \\ 1 & 1 & 0 \\ -2 & -1 & 0 \end{pmatrix}.$$

Then

$$\hat{A} = P^{-1}AP = \begin{pmatrix} 2 & 0 & 0 \\ 0 & 1 & 1 \\ 0 & 0 & 1 \end{pmatrix},$$

and our equation becomes

$$4\hat{x}' = \hat{A}\hat{x} + 2^{-2n}\hat{p}(n), \quad \hat{p}(n) = P^{-1}p(n) = \begin{pmatrix} 0 \\ 0 \\ n+1 \end{pmatrix}.$$

Substituting $\hat{x}(n) = 2^{-2n} \sum_{j=0}^{3} \binom{n}{j} \hat{c}^j$ into equation (12.15), we obtain

$$\Delta\left(\sum_{j=0}^{3} \binom{n}{j}\hat{c}^j\right) = \sum_{j=1}^{3} \binom{n}{j-1}\hat{c}^j = \sum_{j=0}^{3} \binom{n}{j}(\hat{A}-I)\hat{c}^j + p(n),$$

and

$$(\hat{A}-I)\hat{c}^3 = 0$$
$$(\hat{A}-I)\hat{c}^2 = \hat{c}^3$$
$$(\hat{A}-I)\hat{c}^1 = \hat{c}^2 - \begin{pmatrix} 0 \\ 0 \\ 1 \end{pmatrix}$$
$$(\hat{A}-I)\hat{c}^0 = c_1 - \begin{pmatrix} 0 \\ 0 \\ 1 \end{pmatrix},$$

where

$$\hat{A}-I = \begin{pmatrix} 1 & 0 & 0 \\ 0 & 0 & 1 \\ 0 & 0 & 0 \end{pmatrix}.$$

We obtain immediately and in succession:

$$\hat{c}^3 = \alpha_1\begin{pmatrix} 0 \\ 1 \\ 0 \end{pmatrix}, \quad \hat{c}^2 = \alpha_1\begin{pmatrix} 0 \\ 0 \\ 1 \end{pmatrix} + \alpha_2\begin{pmatrix} 0 \\ 1 \\ 0 \end{pmatrix};$$

$$\alpha_1 = 1, \quad \hat{c}^1 = \alpha_2\begin{pmatrix} 0 \\ 0 \\ 1 \end{pmatrix} + \alpha_3\begin{pmatrix} 0 \\ 1 \\ 0 \end{pmatrix};$$

$$\alpha_2 = 1, \quad \alpha_3 = 0, \quad \hat{c}^0 = 0.$$

Hence $\hat{c}^0 = 0$, $\hat{c}^1 = \begin{pmatrix} 0 \\ 0 \\ 1 \end{pmatrix}$, $\hat{c}^2 = \begin{pmatrix} 0 \\ 1 \\ 1 \end{pmatrix}$, and $\hat{c}^3 = \begin{pmatrix} 0 \\ 1 \\ 0 \end{pmatrix}$; i.e.

$$c^0 = 0, \quad c^1 = v^3, \quad c^2 = v^2 + v^3, \quad c^3 = v^2,$$

and

$$x(n) = 2^{-2n}\left(n\begin{pmatrix} 1 \\ 0 \\ 0 \end{pmatrix} + \frac{1}{2}\,n(n-1)\begin{pmatrix} 1 \\ 1 \\ -1 \end{pmatrix} + \frac{1}{6}\,n(n-1)(n-2)\begin{pmatrix} 0 \\ 1 \\ -1 \end{pmatrix}\right).$$

Since

$$\left(\frac{1}{4}\hat{A}\right)^n = \frac{1}{4^n}\begin{pmatrix} 2^n & 0 & 0 \\ 0 & 1 & n \\ 0 & 0 & 1 \end{pmatrix},$$

the general solution of (12.14) is

$$x(n) = 2^{-n}c_1v^1 + 2^{-2n}((c_2+nc_3)v^2 + c_3v^3)$$

$$+ 2^{-n}\left(\begin{pmatrix} -1 \\ -3 \\ 3 \end{pmatrix} + \frac{1}{2}n\begin{pmatrix} 1 \\ 3 \\ -1 \end{pmatrix} - \frac{1}{4}n(n-1)\begin{pmatrix} 1 \\ 1 \\ 2 \end{pmatrix}\right)$$

$$+ 2^{-n}\left(n\begin{pmatrix} 1 \\ 0 \\ 0 \end{pmatrix} + \frac{1}{2}n(n-1)\begin{pmatrix} 1 \\ 1 \\ -1 \end{pmatrix}\right.$$

$$\left. + \frac{1}{6}n(n-1)(n-2)\begin{pmatrix} 0 \\ 1 \\ -1 \end{pmatrix}\right),$$

where

$$v^1 = \begin{pmatrix} 1 \\ 1 \\ -2 \end{pmatrix}, \quad v^2 = \begin{pmatrix} 0 \\ 1 \\ -1 \end{pmatrix} \quad \text{and} \quad v^3 = \begin{pmatrix} 1 \\ 0 \\ 0 \end{pmatrix}.$$

12.12. Exercise. Find the general solution of

$$x_1' = x_1 - x_2 + 4x_3 - 3^n + \cos n\omega$$

$$x_2' = 3x_1 + 2x_2 - x_3 + 3^{n-1} - \sin n\omega$$

$$x_3' = 2x_1 + 2x_2 - x_3 + 3^{n-2}.$$

13. Forced oscillations.

For $f\colon J_0 \to \mathbf{C}^m$, we say that f is *periodic* if, for some positive integer τ,

$$f(n+\tau) = f(n) \quad \text{for all} \quad n \in J_0;$$

τ is called a *period* of f. The least such τ is the least period. If τ is the least period, then the only periods of f are integral multiples of τ. For instance, $e^{i\frac{2\pi\sigma n}{\tau}}$, σ a nonnegative integer, is periodic of period τ. If $(\sigma,\tau) = 1$, τ is the least period. The constant functions have period 1.

The problem of forced oscillations is to determine the periodic solutions of

$$x' = Ax + f(n). \tag{13.1}$$

when the forcing term f (the input) is periodic. We know that the equation

$$x' = Ax + b\cos\frac{2\pi n}{\tau}, \quad b \in \mathbf{R}^m, \tag{13.2}$$

has a periodic solution of period τ if $e^{i\frac{2\pi}{\tau}}$ is not an eigenvalue of A. This periodic solution is the real part of $e^{in\omega}(e^{i\omega}I - A)^{-1}b$ $(\omega = \frac{2\pi}{\tau})$, and, if the free equation $x' = Ax$ has no nontrivial periodic solutions of period τ, this is the only periodic solution of period τ. This condition is always satisfied if A is stable, and every solution approaches this periodic solution as $n \to \infty$. It is therefore called the *steady-state solution*.

Let us look now at the general system (13.1) where f is any periodic function of period τ. We note first that if

(13.1) has a periodic solution $x(n)$ of period τ_0, then $f(n) = x(n) - Ax(n)$, and τ_0 is a period of f. Thus, if τ is the least period of f, it is easy to see that $\tau_0 = k\tau$, k a positive integer. Another simple observation is that, *if $x(n)$ is any solution of (13.1), then $x(n+\tau)$ is also a solution, and $x(n)$ is periodic of period τ if and only if $x(0) = x(\tau)$.* This is a simple consequence of uniqueness of solutions.

The general solution of (13.1) is (equation (12.5)

$$x(n) = A^n c + \sum_{k=0}^{n-1} A^{n-k-1} f(k). \tag{13.3}$$

Thus, $c = x(0) = x(\tau)$ if and only if

$$(I - A^\tau)c = \sum_{k=0}^{\tau-1} A^{n-k-1} f(k), \tag{13.4}$$

and equation (13.1) has a unique periodic solution of period τ if and only if $(I - A^\tau)$ is nonsingular. Hence, we have

13.1. *Proposition*. If f is periodic of period τ, then equation (13.1) has a unique periodic solution of period τ if and only if $(I - A^\tau)$ is nonsingular.

In connection with this result let us now show that

13.2. *Proposition*. Let f be periodic of period τ. Then the following are equivalent:

(i) The free equation $x' = Ax$ has a nontrivial periodic solution of period τ.

(ii) $e^{i\frac{2k\pi}{\tau}} \in \sigma(A)$ for some integer $0 \leq k < \tau$.

(iii) $I - A^\tau$ is singular $(1 \in \sigma(A^\tau))$.

Proof: (i) ⇒ (ii): Assume that $x(n)$ is a solution of $x' = Ax$, $x(0) \neq 0$, and $x(n+\tau) = x(n)$ for all n. Then from what we know about the general solution

$$x(n) = c_1\lambda_1^n v^1 + \cdots + c_s\lambda_s^n v^s,$$

where the λ_i are eigenvalues of A, v^i are associated eigenvectors, and $v^1,\ldots,v^s$ are linearly independent. Since at least one c_i is nonzero, we may assume $c_1 \neq 0$. Then $x(0) = x(\tau)$ implies $\lambda_1^\tau = 1$ or $\lambda_1 = e^{i\,2nk\pi/\tau}$, a $\tau^{\underline{th}}$ root of unity.

(ii) ⇒ (iii). This follows since $\lambda \in \sigma(A)$ implies $\lambda^\tau \in \sigma(A^\tau)$.

(iii) ⇒ (i). $(I - A^\tau)$ singular implies there is a $c \neq 0$ satisfying $(I - A^\tau)c = 0$. Then $x(n) = A^n c$ is a nontrivial periodic solution of the free equation of period τ. □

Putting the above two propositions together we have

13.3. Theorem. Let f be periodic of period τ. Then the following are equivalent:

(i) The equation $x' = Ax + f(n)$ has a unique periodic solution of period τ.

(ii) $I - A^\tau$ is nonsingular $(1 \notin \sigma(A^\tau))$.

(iii) The free equation $x' = Ax$ has no nontrivial solution of period τ.

(iv) $e^{i\,2k\pi/\tau}$ is not an eigenvalue of A for $k = 0,1,\ldots,\tau-1$.

13.4. Corollary (Steady-state solution). Assume that f is periodic of period τ and that A is stable. Then

$x' = Ax + f(n)$ has a unique periodic solution, its period is τ, and all solutions approach this periodic solution as $n \to \infty$.

13.5. Exercise. Let $f(t) = \sum_{k=-\sigma}^{\sigma} b^k e^{ik\omega t}$, $t \in R$, be a trigonometric polynomial, where ω is any nonnegative real number and $b^k \in C^k$. As a function on R, the period of $f(t)$ is $\frac{2\pi}{\omega}$. As a function $f(n)$ on J_0, $f(n)$ is periodic if and only if ω is a rational multiple of 2π. Also for $f(n)$ we can always assume $0 \leq \omega < 2\pi$. If $\frac{\omega}{2\pi} = \frac{\sigma}{\tau}$ and σ and τ are relatively prime, then τ is the least period of $f(n)$.

a. Discuss the solutions of

$$x' = Ax + f(n) \tag{13.5}$$

when f is a trigonometric polynomial.

b. Discuss the solutions of (13.5) when f is a trigonometric (Fourier) series; i.e., $\sigma \to \infty$ with a suitable convergence for the series.

14. Systems of higher order equations $P(z)y = 0$. The equivalence of polynomial matrices.

Let us see by way of an example how we can solve a system of higher order difference equations. Consider

$$\begin{aligned} &y_2'' + y_1' + y_2' + y_3' - y_1 - 3y_2 - y_3 = 0 \\ &y_1''' - y_1'' + y_3'' - 4y_1' - y_3' + 2y_1 = 0 \\ &y_1' + y_2' - y_1 - y_3 = 0. \end{aligned} \tag{14.1}$$

Here we have three linear difference equations of order 3 (z^3 is the higher power of z) and 3 unknown functions. We will find the general solution by removing the coupling between equations by carrying out elementary operations on the equations that carry us from one equivalent system to another. This is similar to Gaussian elimination for solving linear algebraic equations. To do this systematically and to see how to discuss the general case let us rewrite the equations:

$$(z-1)y_1 + (z^2+z-3)y_2 + (z-1)y_3 = 0$$

$$(z^3+z^2-4z+2)y_1 + (z^2-z)y_3 = 0$$

$$(z-1)y_1 \quad + (z-1)y_2 = 0$$

and the problem which now looks easier, is to solve

$$P(z)y = 0$$

where

$$P(z) = \begin{pmatrix} z-1 & z^2+z-3 & z-1 \\ z^3+z^2-4z+2 & 0 & z^2-z \\ z-1 & 0 & z-1 \end{pmatrix}.$$

We carry out the elementary operations on the rows and columns of the matrix and see that they lead from one set of equivalent equations (matrices) to another.

<u>Step 1</u>. Subtract the $3^{\underline{rd}}$ row from the $1^{\underline{st}}$ row. This gives

$$P_1(z) = \begin{pmatrix} 0 & z^2+z-3 & 0 \\ z^3+z^2-4z+2 & 0 & z^2-z \\ z-1 & 0 & z-1 \end{pmatrix}$$

and corresponds to left multiplication of $P(z)$ by

$$L_1 = \begin{pmatrix} 1 & 0 & -1 \\ 0 & 1 & 0 \\ 0 & 0 & 1 \end{pmatrix} \quad \text{and} \quad P_1(z) = L_1 P(z).$$

Step 2. Subtract z times the 3rd row from the second row.

$$P_2(z) = \begin{pmatrix} 0 & z^2+z-3 & 0 \\ z^3-3z+2 & 0 & 0 \\ z-1 & 0 & z-1 \end{pmatrix},$$

$$L_2(z) = \begin{pmatrix} 1 & 0 & 0 \\ 0 & 1 & -z \\ 0 & 0 & 1 \end{pmatrix}, \quad P_2(z) = L_2(z)P_1(z).$$

Step 3. Subtract the 3rd column from the 1st column.

$$P_3(z) = \begin{pmatrix} 0 & z^2+z-3 & 0 \\ z^3-3z+2 & 0 & 0 \\ 0 & 0 & z-1 \end{pmatrix},$$

$$R_1 = \begin{pmatrix} 1 & 0 & 0 \\ 0 & 1 & 0 \\ -1 & 0 & 1 \end{pmatrix}, \quad P_3(z) = P_2(z)R_1.$$

Step 4. Interchange the 1st and 2nd rows.

$$P_4(z) = \begin{pmatrix} z^3-3z+2 & 0 & 0 \\ 0 & z^2+z-3 & 0 \\ 0 & 0 & z-1 \end{pmatrix}$$

$$= \begin{pmatrix} (z-1)^2(z+2) & 0 & 0 \\ 0 & (z-1)(z+2) & 0 \\ 0 & 0 & z-1 \end{pmatrix}$$

$$L_3 = \begin{bmatrix} 0 & 1 & 0 \\ 1 & 0 & 0 \\ 0 & 0 & 1 \end{bmatrix}, \quad P_4(z) = L_3 P_3(z);$$

$$P_4(z) = L(z)P(z)R(z),$$

where $L(z) = L_3 L_2(z) L_1$ and $R(z) = R_1$.

Hence, if we multiply (left multiplication) our original equation by $L(z)$ and place $y = R\overline{y}$, we obtain the decoupled system $\overline{P}(z)\overline{y} = 0$ $(\overline{P}(z) = P_4(z))$; i.e.,

$$\begin{aligned} (z-1)^2(z+2)\overline{y}_1 &= 0 \\ (z-1)(z+2)\overline{y}_2 &= 0 \\ (z-1)\overline{y}_3 &= 0, \end{aligned} \tag{14.2}$$

where

$$\begin{aligned} y_1 &= \overline{y}_1 & \qquad \overline{y}_1 &= y_1 \\ y_2 &= \overline{y}_2 \quad , & \overline{y}_2 &= y_2 \\ y_3 &= -\overline{y}_1 + \overline{y}_3 & \overline{y}_3 &= y_1 + y_3. \end{aligned}$$

Each of the right and left elementary matrices have the property that they are nonsingular and their determinants are constants (they do not depend on z); $L^{-1}(z)$ and $R^{-1}(z)$ are therefore polynomial matrices (the elements of the matrices and polynomials). The steps are all reversible and (14.1) and (14.2) are equivalent.

Solving (13.2), we obtain

$$\begin{aligned} y_1(n) &= \overline{y}_1(n) = (c_{11} + c_{12}n) + c_{13}(-2)^n \\ y_2(n) &= \overline{y}_2(n) = c_{21} + c_{22}(-2)^n \\ y_3(n) &= -\overline{y}_1(n) + \overline{y}_3(n) = -(c_{11}+c_{12}n) - c_{13}(-2)^n - c_{33}. \end{aligned}$$

There are 6 arbitrary constants in the general solution. The state of (14.2) is

$$\begin{pmatrix} \bar{y}_1 \\ \bar{y}_1' \\ \bar{y}_1'' \\ \bar{y}_2 \\ \bar{y}_2' \\ y_3 \end{pmatrix}, \qquad \text{and} \qquad \begin{pmatrix} y_1 \\ y_1' \\ y_1'' \\ y_2 \\ y_2' \\ y_3 \end{pmatrix}$$

is the state of (14.1). We can expect (14.1) to be equivalent to a $6^{\underline{th}}$ order system $x' = Ax$ and, of course, it is. In terms of companion matrices

$$A = \text{diagonal}\left\{ \begin{pmatrix} 0 & 1 & 0 \\ 0 & 0 & 1 \\ -2 & 3 & 0 \end{pmatrix}, \begin{pmatrix} 0 & 1 \\ 3 & -1 \end{pmatrix}, 1 \right\}.$$

The matrix $P_4(z)$ is a canonical form for $P(z)$ (the equations (14.2) are a canonical or normal form for (14.1) and the coordinates $\bar{y}$ are canonical or normal coordinates). The polynomials $(\lambda-1)^2(\lambda+2)$, $(\lambda-1)(\lambda+2)$, and $\lambda-1$ are called the "invariant polynomials" of $P(z)$ -- the second divides the first, the first divides the second. The matrices along the diagonal of A are the principal companion matrices of the invariant polynomials of $P(z)$.

14.1. Exercise. The elementary divisors of A are the elementary divisors of $P(z)$. Show that: (a) The elementary divisors of $P(z)$ are $(z-1)^2$, $(z-1)$, $(z-1)(z+2)$, and $(z+2)$. (b) The minimal polynomial of A is the first elementary divisor of $P(z)$.

What was done in the above example can be generalized.

The difficulty lies in expressing everything precisely. We shall not attempt to do that here but will simply state the principal result. A good reference is Gautmacher [1], Vol. 1, Chapter VI, "Equivalent transformations of polynomial matrices. Analytic theory of elementary divisors."

An $r \times s$ matrix $P(\lambda) = (p_{ij}(\lambda))$, each $p_{ij}(\lambda)$ a polynomial in λ, is called a <u>polynomial matrix</u>. The <u>Rank</u> k of $P(\lambda)$ is the maximum over all λ_0 of the rank of $P(\lambda_0)$ for each fixed λ_0; in other words, if $k > 0$, the matrix has minors of order k not identically equal to zero and all minors of order greater than k are identically equal to zero. Systems of higher order linear difference equations are of the form

$$P(z)y = 0.$$

Two $r \times s$ polynomial matrices $P(\lambda)$ and $Q(\lambda)$ are said to be <u>equivalent</u> if $Q(\lambda) = L(\lambda)Q(\lambda)R(\lambda)$, where $P(\lambda)$ and $Q(\lambda)$ are $r \times r$ and $s \times s$ matrices, respectively, with constant nonzero determinants; i.e. $P^{-1}(\lambda)$ and $Q^{-1}(\lambda)$ exist for each λ and are polynomial matrices.

14.2. <u>Proposition</u>. Each polynomial matrix $P(\lambda)$ is equivalent to a canonical diagonal matrix

$$\begin{pmatrix} \alpha_k(\lambda) & 0 & \cdot & \cdot & \cdot & \cdot & 0 \\ 0 & \alpha_{k-1}(\lambda) & & & & & \cdot \\ & & \ddots & & & & \cdot \\ \cdot & & & \alpha_1(\lambda) & & & \cdot \\ \cdot & & & & \ddots & & \cdot \\ \cdot & & & & 0 & & 0 \\ 0 & \cdot & \cdot & \cdot & \cdot & 0 & 0 \end{pmatrix},$$

where k is the Rank of $P(\lambda)$, each $\alpha_j(\lambda)$ is a monic polynomial (not identically equal to zero), and each $\alpha_j(\lambda)$ divides α_{j+1}, $j = 1,\ldots,r-1$.

The polynomial $\alpha_k(\lambda),\ldots,\alpha_1(\lambda)$ are called the _invariant polynomials_ of $P(\lambda)$. It can be shown that _two polynomial matrices are equivalent if and only if they have the same canonical forms_ (i.e., the same invariant polynomials), and the invariant polynomials are a complete set of invariants. Note it can be that some of the $\alpha_j(\lambda)$ are identically equal to 1. Of course, if $\alpha_j = 1$, then $\alpha_i = 1$ for $i = 1,\ldots,j$.

For A an $m \times m$ matrix, the above result can be applied to the characteristic matrix $P(\lambda) = \lambda I - A$ to obtain the Jordan canonical form theorem, and from this result one can derive algorithms for computing the Jordan canonical form of a matrix and the transformation of coordinates (the matrix P in Proposition 8.1) that leads to the canonical form (see the above reference to Gantmacher [1]). An important result here is that _two square matrices_ A _and_ B _are similar if and only if the matrix polynomials_ $\lambda I - A$ _and_ $\lambda I - B$ _are equivalent_. This is to be expected since the system $x' = Ax$ is a system $(zI-A)x = 0$ of first order equations, and the Rank of $(\lambda I-A)$ is m.

Proposition 14.2 can then be used to discuss higher order systems of equations

$$P(z)y = 0.$$

15. The control of linear systems. Controllability.

Within control and system theory the fundamentally important concept of controllability arose naturally during the early development of optimal control theory in the late 1950's and was discovered independently by a number of mathematicians and engineers in the United States and the Soviet Union.

Let us consider

$$x' = Ax + f(n). \tag{15.1}$$

The system to be controlled is $x' = Ax$ (see Figure 15.1) and (15.1) is the control system. The input $f(n)$ is the control function

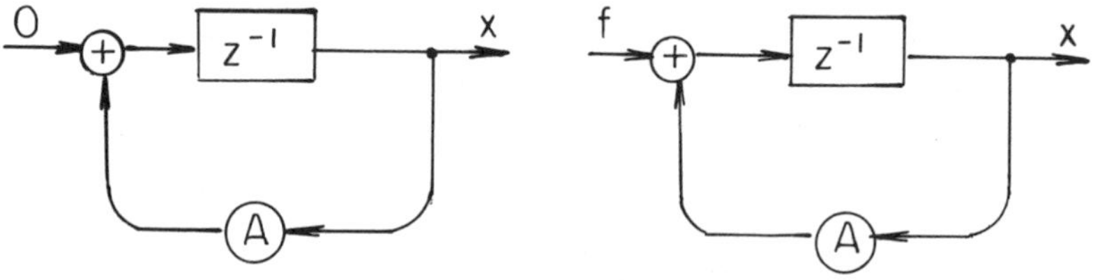

(a) The uncontrolled system

(b) The controlled system

Figure 15.1

Man is faced with many different types of control problems -- landing a vehicle on Mars, controlling the economy, illnesses, epidemics, populations, floods, crime, manufacturing processes, etc. We can identify two broad classes of these. We may wish starting at some initial state x^0 to bring the system to some desired state y within or at some given time n_1. This is, for instance, the problem of landing a vehicle on Mars or bringing the economy to a given state before a presidential election or reducing a patient's blood pressure. The problem is to hit a given target within or at a given time and is called by engineers the <u>ballistic problem</u>. Another type of problem is to have the output $x(n)$ follow a desired trajectory $w(n)$ -- one would like to select $f(n)$ so that over a period of time $x(n) = w(n)$. Engineers call this problem the <u>servomechanism problem</u>. Taken literally, the servomechanism problem requires, given x^0, to bring the system to $x^1 = w(1)$ in unit time, then from x^1 to $x^2 = w(2)$ in unit time, and so forth. This requires bringing x^0 to y in unit time for arbitrary x^0 and y. With no constraints on the control $f(n)$, this can always be done since

$$x(1) = Ax(0) + f(0).$$

This is both trivial and unrealistic. It assumes, for instance, in the way in which the control appears, that the control force can be applied to affect directly each of the state variables. An example where this cannot be done is the second order equation

$$y'' + a_1 y' + a_0 y = g(n).$$

An equivalent system (Figure 15.2) is

$$x_1' = x_2$$
$$x_2' = -a_0 x_1 - a_1 x_2 + g(n);$$

$x_1 = y$, $x_2 = y'$.

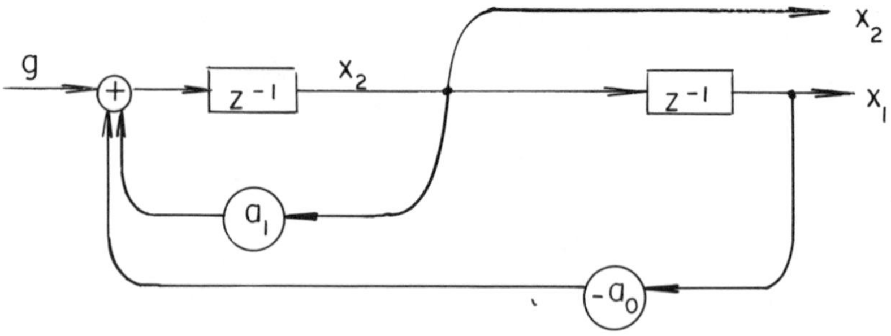

Figure 15.2

There is only one component of control -- $f(n) = \begin{bmatrix} 0 \\ g(n) \end{bmatrix}$. At time n+1

$$x_1(n+1) = x_2(n)$$
$$x_2(n+1) = -a_0x_1(n) - a_1x_2(n) + g(n).$$

The control g at time n affects x_2 at time n+1 directly but not x_1. Its affect on x_1 is through its affect on $x_2(n)$; i.e., through dynamic coupling. Thus, $y = x_1$ might be blood pressure, and g the affect of a drug -- g(0), g(1), etc. a schedule for taking the drug. The control g(n) affects directly only $x_2'(n) = y''(n) = y(n+2)$ so that eventually it affects the blood pressure. This is like braking an automobile, which affects directly only the acceleration. We would expect this model to say the blood pressure can be controlled, and, as we shall see in a moment, it is controllable. Similarly, in controlling an economy or an epidemic we would hardly expect to be able to control directly all of the state variables.

A more realistic model (see Figure 15.3) is

$$x' = Ax + u_1(n)b^1 + \cdots + u_r(n)b^r,$$

where there are r control variables $u_j(n)$. Thus (Figure 15.3)

$$x' = Ax + Bu(n), \tag{15.2}$$

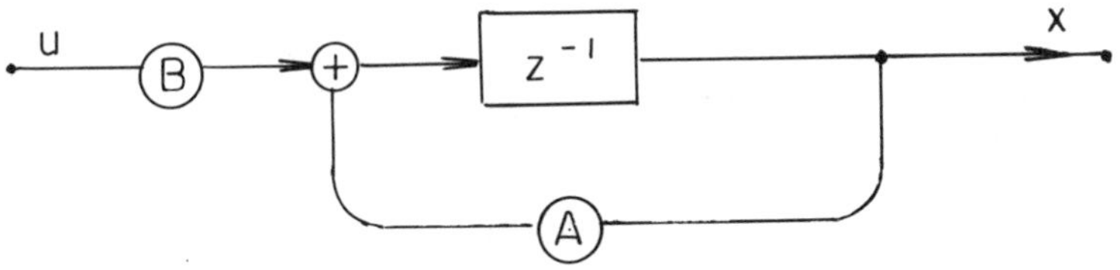

Figure 15.3

where A is an $m \times m$ matrix, $B = (b^1,\ldots,b^r)$ is an $m \times r$ matrix, x is an m-vector and u is an r-vector. We assume at first no constraints on $u(n)$ and ask if it is possible, given an initial state x and a desired state y, to reach y in time n.

For each control function $u\colon J_0 \to \mathbf{R}^r$ and $x \in \mathbf{R}^m$ let $\pi(n,u,x)$ denote the solution of (15.2) satisfying $x(o,u,x) = x$. By the variation of constants formula

$$\pi(n,u,x) = A^n x + \sum_{j=0}^{n-1} A^{n-j-1}Bu(j)$$

or

$$\pi(n,u,x) = A^n x + \sum_{j=0}^{n-1} A^j Bu(n-j-1).$$

We will say that y is <u>reachable</u> <u>from</u> x <u>in</u> <u>time</u> n if $y = \pi(n,u,x)$ for some control u. If for some integer n each $y \in \mathbf{R}^m$ is reachable from each $x \in \mathbf{R}^m$ in time n, the system $x' = Ax + Bu$, or the pair $\{A,B\}$, is said to be <u>controllable</u> (<u>in time</u> n).

Since

$$\pi(1,u,x) = Ax + Bu(0),$$

we see that (15.2) is controllable in time 1 if and only if B is $m \times m$ and nonsingular. (This is equivalent to (15.1) (make the change of coordinates $x = B\hat{x}$). A state y is reachable from x in time 2 if and only if

$$\begin{aligned} y - A^2x &= Bu(1) + ABu(0) \\ &= (B,AB)\begin{pmatrix} u(1) \\ u(0) \end{pmatrix} \\ &= Y(2)\hat{u}_2, \end{aligned}$$

where $Y(2) = (B,AB)$ is an $m \times 2r$ matrix and $\hat{u}_2 = \begin{pmatrix} u(1) \\ u(0) \end{pmatrix}$ is a $2r$-dimensional vector; $Y(2)$ is a linear transformation of $\mathbf{R}^{2r}$ into $\mathbf{R}^m$. Defining

$$Y(n) = (B,AB,\ldots,A^{n-1}B)$$

and

$$\hat{u}(n) = \begin{pmatrix} u(n-1) \\ \vdots \\ u(0) \end{pmatrix},$$

we have

$$\pi(n,u,x) = A^n x + Y(n)\hat{u}(n), \tag{15.3}$$

and y is reachable from x in time n if and only if for some control u

$$y - A^n x = Y(n)\hat{u}(n);$$

$Y(n)$ is an $m \times rn$ matrix and $\hat{u}(n)$ is an rn-dimensional vector. Hence, we see that

15.1. Proposition. A state y is reachable from x in time n if and only if $y - A^n x \in \text{image } Y(n)$. The system $x' = Ax + Bu$ is controllable if and only if for some n image $Y(n) = \mathbf{R}^m$; i.e., if and only if $\text{rank } Y(n) = m$ for some n.

15.2. Example. a. Consider

$$y'' + a_1 y' + a_0 y = u(n).$$

An equivalent system is

$$x_1' = x_2$$
$$x_2' = -a_0 x_1 - a_1 x_2 + u(n);$$

$$A = \begin{bmatrix} 0 & 1 \\ -a_0 & -a_1 \end{bmatrix}, \quad B = b = \begin{bmatrix} 0 \\ 1 \end{bmatrix},$$

and

$$Ab = \begin{bmatrix} 1 \\ -a_1 \end{bmatrix}.$$

The vectors $\begin{bmatrix} 0 \\ 1 \end{bmatrix}$ and $\begin{bmatrix} 1 \\ -a_2 \end{bmatrix}$ are linearly independent, and $\text{rank } Y(2) = \text{rank}\begin{bmatrix} 0 & 1 \\ 1 & -a_2 \end{bmatrix} = 2$. A second-order equation is always controllable in time 2; i.e., it is always controllable.

b. Consider

$$x_1' = a_{11}x_1 + a_{12}x_2$$

$$x_2' = a_{21}x_1 + a_{22}x_2 + u(n).$$

Here $A = \begin{pmatrix} a_{11} & a_{12} \\ a_{21} & a_{22} \end{pmatrix}$, $B = b = \begin{pmatrix} 0 \\ 1 \end{pmatrix}$, and $Ab = \begin{pmatrix} a_{12} \\ a_{22} \end{pmatrix}$.

Thus the system is controllable if $a_{12} \neq 0$ (there is dynamic coupling between x_1 and x_2). If $a_{12} = 0$, then $x_1(n) = (a_{11})^n x_1(0)$, and the system is not controllable.

c. Consider

$$x_1' = a_{11}x_1 + a_{12}x_2 + u(n)$$

$$x_2' = a_{21}x_1 + a_{22}x_2 + u(n).$$

Here $B = b = \begin{pmatrix} 1 \\ 1 \end{pmatrix}$ and $Ab = \begin{pmatrix} a_{11} + a_{12} \\ a_{21} + a_{22} \end{pmatrix}$, and the system is controllable if $a_{11} + a_{12} \neq a_{21} + a_{22}$ -- b and Ab are linearly independent. We will see in a moment that this condition is also necessary.

15.3. <u>Exercise</u>. Obtain a necessary and sufficient condition for

$$x_1' = a_{11}x_1 + a_{12}x_2 + u(n)$$

$$x_2' = a_{21}x_1 + a_{22}x_2 - u(n)$$

to be controllable.

We want to show now that in determining whether or not rank $Y(n) = m$ for some n we need only look at $n \leq m$; i.e., if the system is not controllable in time m, the dimension of the state space, it is never controllable. Note that

$$Y(1) = B = (b^1,\ldots,b^r)$$

$$Y(2) = (B,AB) = (b^1,\ldots,b^r,Ab^1,\ldots,Ab^r)$$

and

$$Y(n+1) = (Y(n), A^nB). \tag{15.4}$$

We show first a partial result.

15.4. Proposition. The following are equivalent:

(i) rank $Y(n)$ = rank $Y(n+1)$

(ii) image $Y(n)$ = image $Y(n+1)$.

(iii) image $(A^nB) \subset$ image $Y(n)$

(iv) rank $Y(n)$ = rank $Y(n+j)$, $j \geq 0$.

Proof: (i) $\Rightarrow$ (ii): This is a consequence of the fact that image $Y(n) \subset$ image $Y(n+1)$.

(ii) $\Rightarrow$ (iii): Follows immediately from (15.4).

(iii) $\Rightarrow$ (i): We see from (15.4) that (iii) $\Rightarrow$ (ii) and obviously (ii) $\Rightarrow$ (i). (i) $\Leftrightarrow$ (iv): We now know that (i), (ii) and (iii) are equivalent. Since A image (A^nB) = image $(A^{n+1}B)$ and $AY(n)$ image $Y(n+1)$, we see that (iii) implies image $(A^{n+1}B)$ image $Y(n+1)$, and this implies rank $Y(n+1)$ = rank $Y(n+2)$. Hence (i) $\Rightarrow$ (iv), and obviously (iv) $\Rightarrow$ (i). □

From Proposition 15.1 and 15.4 we can now give a characterization of controllability. We show first

15.5. Proposition. Let s be the degree of the minimum polynomial of A $(s \leq m)$. There is an integer $k \leq s$ such that rank $Y(1) <$ rank $Y(2) < \cdots <$ rank $Y(k)$ = rank $Y(k+j)$, all $j \geq 0$.

Proof: From Proposition 15.4 we know that such an integer k exists since rank $Y(n) \leq m$ for all n, and hence $k \leq m$. Let $\psi(\lambda) = \lambda^s + a_{s-1}\lambda^{s-1} + \cdots + a_0$ be the minimal polynomial of A. Then $\psi(A)B = 0$, and image $(A^s B) \subset$ image $Y(s)$. It then follows from Proposition 15.4 that $k \leq s$. □

This proposition tells us that image $Y(n)$ is strictly increasing (image $Y(n-1)$ is properly contained in image $Y(n)$) for $0 \leq n \leq k$ and image $Y(n)$ is constant for $n \geq k$. We then have immediately from Proposition 15.1

15.6. Theorem. The system $x' = Ax + Bu$ is controllable if and only if $\operatorname{rank} Y(k) = \operatorname{rank}(B, AB, \ldots, A^{k-1}B) = m$ for some integer $k \leq s$, where s is the degree of the minimal polynomial of A.

We see from this theorem that, if $x' = Ax + Bu$ is controllable, then every state y can be reached from each x in time s, and therefore in time m.

15.7. Example. Consider

$$x_1' = x_2 + u_1(n) + u_2(n)$$

$$x_2' = x_3 + u_1(n) - u_2(n)$$

$$x_3' = u_1(n).$$

Here $A = \begin{pmatrix} 0 & 1 & 0 \\ 0 & 0 & 1 \\ 0 & 0 & 0 \end{pmatrix}$ and $B = \begin{pmatrix} 1 & 1 \\ 1 & -1 \\ 1 & 0 \end{pmatrix}$; $A^3 = 0$ and the minimal polynomial is λ^3 $(s = m = 3)$.

$$AB = \begin{pmatrix} 1 & -1 \\ 1 & 0 \\ 0 & 0 \end{pmatrix} \quad \text{and} \quad Y(2) = \begin{pmatrix} 1 & 1 & 1 & -1 \\ 1 & -1 & 1 & 0 \\ 1 & 0 & 0 & 0 \end{pmatrix}.$$

Since $\operatorname{rank} Y(2) = 3$, the system is controllable in time 2. Also, since $b^1 = \begin{pmatrix} 1 \\ 1 \\ 1 \end{pmatrix}$, $Ab^1 = \begin{pmatrix} 1 \\ 1 \\ 0 \end{pmatrix}$ and $A^2 b^1 = \begin{pmatrix} 1 \\ 0 \\ 0 \end{pmatrix}$, we see that the system is controllable with $u_2 = 0$ but in time 3. Since $b^2 = \begin{bmatrix} 1 \\ -1 \\ 0 \end{bmatrix}$, $Ab^2 = \begin{bmatrix} -1 \\ 0 \\ 0 \end{bmatrix}$, and $A^2b^2 = \begin{bmatrix} 0 \\ 0 \\ 0 \end{bmatrix}$, the system is not controllable with $u_1 = 0$. The control u_2 does, however, contribute to the speed of control.

Let us look now at two other characterizations of controllability. From equation (15.3) we have

$$\pi(n,u,x) = A^n x + Y(n)\hat{u}_n,$$

and we see that if there is a nonzero vector $v \in \mathbf{R}^m$ such that $v^T Y(n) = 0$ for all $n \geq 0$ then

$$v^T \pi(n,u,x) = v^T A^n x \quad \text{for all} \quad n \geq 0.$$

and the system is not controllable -- the control $u(n)$ has no affect on the component of $\pi(n,u,x)$ in the direction of v. It is also clear that $v^T Y(n) = 0$ for all $n \geq 0$ and $v \neq 0$ implies $\operatorname{rank} Y(n) < m$ for all $n \geq 0$. Conversely, if $\operatorname{rank} Y(n) = m$ for some n there can be no such v. Hence the system $x' = Ax + Bu$ is controllable if and only if $v^T Y(n) = 0$ for all $n \geq 0$ implies $v = 0$.

Another characterization can be given in terms of the $m \times m$ matrix

$$W(n) = Y(n)Y^T(n) = \sum_{j=0}^{n-1} A^j BB^T (A^j)^T. \tag{15.5}$$

The matrix $W(n)$ is symmetric and positive semidefinite, and it is clear that the system is controllable if and only if $W(m)$ is positive definite, which is equivalent to $W(n)$ positive definite for all $n \geq m$. Hence we have

15.8. <u>Theorem</u>. The following are equivalent:

(i) The system $x' = Ax + Bu$ (or $\{A,B\}$) is controllable.

(ii) rank $Y(m) = m$.

(iii) $v^T Y(n) = 0$ for all $n \geq 0$ implies $v = 0$.

(iv) $W(m)$ is positive definite.

The state of an $m^{\underline{th}}$-order equation

$$\psi(z)y = y^{(m)} + a_{m-1}y^{(m-1)} + \cdots + a_0 y = u(n) \tag{15.6}$$

is

$$\bar{x} = \begin{pmatrix} y \\ y' \\ \vdots \\ y^{(m-1)} \end{pmatrix},$$

and we know that (15.6) is equivalent to the system

$$\bar{x}' = A_0\bar{x} + \delta^m u(n) \tag{15.7}$$

where

$$A_0 = \begin{pmatrix} 0 & 1 & 0 & \cdots & 0 \\ \vdots & & \ddots & \ddots & \vdots \\ \vdots & & & \ddots & 0 \\ 0 & & & 0 & 1 \\ -a_0 & -a_1 & \cdots & & -a_m \end{pmatrix} \quad \text{and} \quad \delta^m = \begin{pmatrix} 0 \\ \vdots \\ 0 \\ 1 \end{pmatrix},$$

A_0 is the principal companion matrix of $\psi(\lambda)$. Thus, controllability of (15.6) means controllability of the state $\bar{x}$ and therefore controllability of (15.7). Now

$$\delta^m = \begin{pmatrix} 0 \\ \vdots \\ \vdots \\ 0 \\ 1 \end{pmatrix}, \quad A_0\delta^m = \delta^{m-1} = \begin{pmatrix} 0 \\ \vdots \\ \vdots \\ 0 \\ 1 \\ 0 \end{pmatrix}, \ldots A_0^{m-1}\delta^m = \delta^1 = \begin{pmatrix} 1 \\ 0 \\ \vdots \\ \vdots \\ 0 \end{pmatrix},$$

and hence we see that <u>every m^{th} order equation</u> $\psi(z)y = u(n)$ <u>is controllable</u>.

Let us look at the general case of one control variable. Then $B = b \in \mathbf{R}^m$, and the system is

$$x' = Ax + bu(n). \tag{15.8}$$

Just as in the proof of Proposition 10.5 we see that, if (15.8) is controllable then it is equivalent to an mth order equation (15.6). Hence we have the following result.

15.9 <u>Proposition</u>. A system $x' = Ax + bu(n)$ is controllable if and only if it is equivalent to an mth-order equation $\psi(z)y = u(n)$.

15.10. <u>Exercise</u>. Show that: A necessary condition that $x' = Ax + bu$ be controllable is that A be a companion matrix.

To the system $x' = Ax + Bu(n)$ we can add, through linear feedback (see Figure 15.4), Cx to the input, and the resulting system is $x' = Ax + B(u(n) + Cx)$, or

$$x' = (A+BC)x + u(n).$$

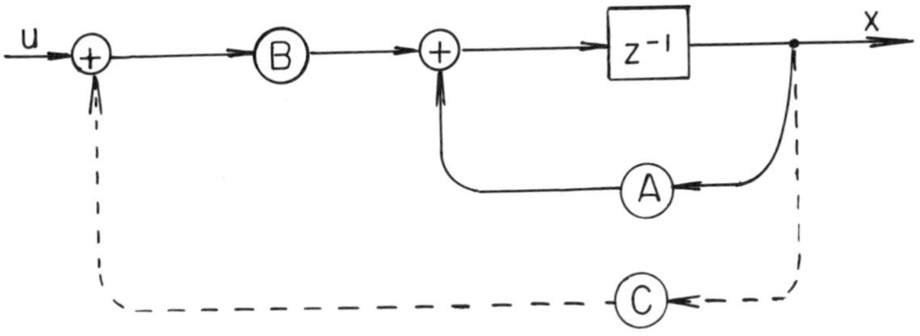

<u>The Addition of Linear Feedback</u>

Figure 15.4

We want to show that this does not affect controllability; controllability is invariant under the addition of linear feedback. Here A is a real $m \times m$ matrix, B is a real $m \times r$ matrix, and C is a real $r \times m$ matrix.

15.11. Proposition. $\{A+BC_0, B\}$ controllable for some C_0 implies $\{A+BC, B\}$ is controllable for all C.

Proof: Assume that $v^T A^n B = 0$ for all $n \geq 0$. Then

$$v^T(A+BC)^n B = v^T(A+BC)(A+BC)^{n-1}B$$

$$= v^T A(A+BC)^{n-1}B = \cdots = v^T A^n B = 0$$

for all $n \geq 0$ and all C. Hence, by Theorem 15.8, $\{A+BC_0, B\}$ controllable for some C_0 implies $\{A,B\}$ is controllable. Hence $\{(A+BC)-BC, B\}$ is controllable for all C, and, by what was just shown, $\{A+BC,B\}$ is controllable for all C. □

We can now reinterpret Proposition 10.4. If the system $x' = Ax + bu$ is controllable, then, through linear feedback $u = c^T x$, the resulting system is $x' = (Ax + bc^T)x$, and the eigenvalues of $(Ax + bc^T)$ (the spectrum $\sigma(A+bc^T)$) can be arbitrarily assigned, and hence the system can be stabilized by linear feedback. This is the problem engineers call "pole assignment" and will be discussed more completely in the next section.

15.12. Exercise. Let $B = \{b^1,\ldots,b^r\} \neq 0$ so that we may assume $b^i \neq 0$, $i = 1,\ldots,r$. Then for each b^i there is a maximum $1 \leq j \leq m$ with the property that b^i, $Ab^i,\ldots,$ $A^{j-1}b^i$ are linearly independent. If the pair $\{A,B\}$ is controllable and $j < m$, show that there is a b^k such that $b^i, Ab^i, \ldots, A^{j-1}b^i, b^k$ are linearly independent.

16. Stabilization by linear feedback. Pole assignment.

We saw at the end of the previous section that, if a linear control system with one control variable

$$x' = Ax + bu \tag{16.1}$$

is controllable, then by the using linear feedback $u = c^T x$ the system becomes

$$x' = (A + bc^T)x; \tag{16.2}$$

in this special case linear feedback can be used to stabilize the system; in fact, by the choice of c^T we have complete control of the spectrum of $(A+bc^T)$ (see Proposition 10.4). Stabilization by linear feedback is the oldest method for the analysis and design of feedback controls and dates back at least to the early part of the 19th century (see Fuller [1]). It was almost the only method used up to the 1950's and remains of importance up to the present time. The result we present here is of more recent origin. It had been looked at by Langenhop [1] in 1964 over the complex field and was discovered independently by a number of engineers (see Wonham [1] and Padulo and Arbib [1, pp. 596-601]).

For the more general system with r control variables

$$x' = Ax + Bu(n) \tag{16.3}$$

the addition of linear feedback Cx (Figure 16.1) gives

$$x' = (A+BC)x + Bu(n). \tag{16.4}$$

We know that (16.4) is controllable if and only if (16.3) is controllable. The question we want to answer is: when do we have complete control of the spectrum of $(A+BC)$? Engineers who think in terms of transpose z-transforms transform functions and call this "pole assignment".

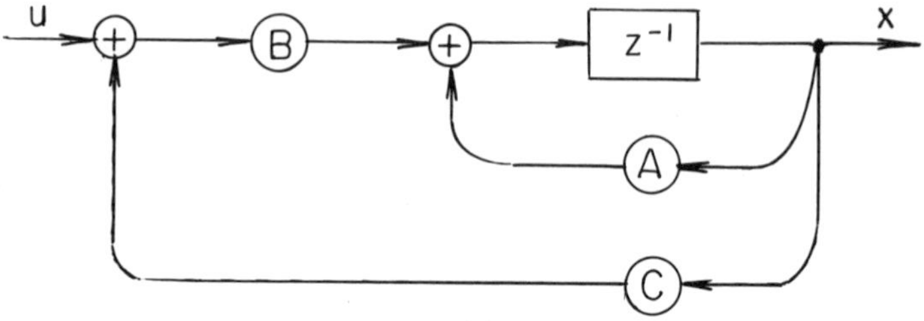

Figure 16.1

Let us look first at the special case $m = 3$ and $r = 2$; i.e.,

$$x' = Ax + Bu = Ax + b^1u_1 + b^2u_2. \tag{16.5}$$

Assume that (16.5) is controllable. If b^1, Ab^1, A^2b^1 are linearly independent, then the system is controllable with $u_2 = 0$, and we know in this case that with feedback control c^Tx we can arbitrarily assign the spectrum of $A + bc^T$. By changing the numbering of b^1 and b^2, if necessary, we know that, if the system is not controllable with one component of control, b^1, Ab^1, A^2b^1 are linearly dependent and b^1, Ab^1, b^2 are linearly independent. Then $A^2b^1 = a_0b_1 + a_1Ab_1$. Make the change of coordinates $x = P\overline{x}$ where $P = (b^1,Ab^1,b^2)$. In the new coordinates

$$\overline{x}' = \overline{A}\,\overline{x} + P^{-1}b^1u_1 + P^{-1}b^2u_2,$$

where

$$\bar{A} = \begin{bmatrix} 0 & a_0 & \alpha_1 \\ 1 & a_1 & \alpha_2 \\ 0 & 0 & \alpha_3 \end{bmatrix},$$

$$P^{-1}b^1 = \begin{pmatrix} 1 \\ 0 \\ 0 \end{pmatrix} = \delta^1 \quad \text{and} \quad P^{-1}b^2 = \begin{pmatrix} 0 \\ 0 \\ 1 \end{pmatrix} = \delta^3;$$

the values of α_1, α_2, α_3 are of no concern -- $Ab^2 = \alpha_1 b^1 + \alpha_2 Ab^1 + \alpha_3 b^2$. Then

$$\bar{x}_1' = a_0\bar{x}_2 + \alpha_1\bar{x}_3 + u_1$$

$$\bar{x}_2' = \bar{x}_1 + a_1\bar{x}_2 + \alpha_2\bar{x}_3$$

$$\bar{x}_3' = \alpha_3\bar{x}_3 + u_2.$$

with

$$u_3 = \bar{c}_3\bar{x}_3, \quad u_1 = \bar{c}_1\bar{x}_1 + \bar{c}_2\bar{x}_2,$$

$$\bar{x}' = (\bar{A} + \bar{B}\,\bar{C})\bar{x} = \hat{A}\bar{x},$$

where

$$\bar{B} = \begin{pmatrix} 1 & 0 \\ 0 & 0 \\ 0 & 1 \end{pmatrix}, \quad \bar{C} = \begin{pmatrix} \bar{c}_1 & \bar{c}_2 & 0 \\ 0 & 0 & \bar{c}_3 \end{pmatrix}$$

and

$$\hat{A} = \begin{bmatrix} \bar{c}_1 & a_0+\bar{c}_2 & \alpha_1 \\ 1 & a_1 & \alpha_2 \\ 0 & 0 & \alpha_3+\bar{c}_3 \end{bmatrix}$$

Then

$$\phi(\lambda) = \det(I-\lambda\hat{A}) = (\lambda-\alpha_3-\bar{c}_3)(\lambda^2-(a_1+\bar{c}_1)\lambda+a_1\bar{c}_1-a_0-\bar{c}_2),$$

and it is clear that by the choice of $\bar{c}_1$, $\bar{c}_2$, and $\bar{c}_3$ we can make the coefficients of the characteristic polynomial anything we please -- with everything real, the coefficients are real.

There is another way of arriving at the same conclusion

and this is the procedure we will follow. In the above add the feedback control

$$\begin{bmatrix} 0 \\ 0 \\ \bar{c}_2\bar{x}_2 \end{bmatrix}$$

to the input. Then

$$\hat{A} = \begin{bmatrix} 0 & a_0 & \alpha_1 \\ 1 & a_1 & \alpha_2 \\ 0 & \bar{c}_2 & \alpha_3 \end{bmatrix}.$$

Now

$$\delta^1 = \begin{bmatrix} 1 \\ 0 \\ 0 \end{bmatrix}, \quad A\delta^1 = \delta^2 = \begin{bmatrix} 0 \\ 1 \\ 0 \end{bmatrix}, \quad \text{and} \quad A^2\delta^1 = \begin{bmatrix} a_0 \\ a_1 \\ \bar{c}_2 \end{bmatrix}.$$

With $\bar{c}_2 \neq 0$, δ^1, $A\delta^1$, $A^2\delta^1$, are linearly independent, and the system

$$\bar{x}' = \hat{A}x' + \delta^1 u^1$$

is controllable; we are back to the case where we know we can control the spectrum. In the original coordinates we have found a matrix C such that

$$x' = (A+BC)x + b^1u^1$$

is controllable. There are, of course, a large number of feedback matrices C that will do this.

16.1. <u>Exercise</u>.

$$\begin{aligned} x_1' &= 2x_1 + 2x_2 + x_3 + u_2 \\ x_2' &= -2x_1 - x_2 \quad\quad + u_1 \\ x_3' &= -2x_1 - x_2 \quad\quad -2u. \end{aligned}$$

Show that: (a) This system is controllable. (b) The uncontrolled system is unstable. Determine a linear feedback control $u(x) = Cx$ that makes the eigenvalues of the controlled system (i) $\{\frac{1}{2}, \frac{1}{3}, \frac{1}{4}\}$; (ii) $\{0,0,0\}$.

16.2. <u>Exercise</u>. The equation

$$x'' + a_1x^1 + a_0x = u(x,x')$$

is equivalent to the system

$$x' = y$$
$$y' = -a_0x - a_1y + u(x,y).$$

We know that we can stabilize the system by linear feedback $u(x,y) = c_1x + c_2y$; in fact, we can make the eigenvalues anything we please. For what values of b and a^2 can you find c_1 and c_2 such that $V(x,y) = x^2 + 2bxy + a^2y$ is positive definite and $\dot{V}(x,y)$ is negative definite. When you do this, what does it tell you about the transient behavior of the solutions as they approach the origin.

We want now to generalize what was done above. Here A is a real $m \times m$ matrix, $B = (b^1,\ldots,b^r)$ is a real $m \times r$ matrix, and C is a real $r \times m$ matrix. We will say that the system $x' = Ax + Bu$, or the pair $\{A,B\}$, is σ-<u>controllable</u> , if, given any set $\sigma_0 = \{\lambda_1,\ldots,\lambda_m\}$ of m complex numbers with the property that $\bar{\sigma}_0 = \{\bar{\lambda}_1,\ldots,\bar{\lambda}_m\} = \sigma_0$ (here the "bar" is the complex conjugate), there is a matrix C such that $\sigma\{A+BC\} = \sigma_0$; in other words, given any real monic polynomial $\phi(\lambda)$ of degree m, there is a matrix C such that the characteristic polynomial of $A+BC$ is $\phi(\lambda)$. We know already that, for $r = 1$, controllability of $\{A,b\}$ implies $\{A,b\}$ σ-controllable.

16.3. <u>Proposition</u>. If the pair $\{A,B\}$ is controllable, then for each column vector b^i of B $(b^i \neq 0)$ there is a matrix C such that $\{A+BC,b^i\}$ is controllable.

<u>Proof</u>: Assume that $\{A,B\}$ is controllable. Let j be the largest integer for which $b^1, Ab^1,\dots,A^{j-1}b^1$ are linearly independent; $1 \leq j \leq m$. Then $A^jb^1 = a_0b^1 + a_1Ab^1 + \cdots + a_{j-1}A^{j-1}b^1$. If $j = m$, we are through, so assume $j < m$. Controllability of $\{A,B\}$ tells us (see Exercise 15.12) that there is a b^i (and we may take $i = 2$) such that $b^1,\dots,A^{j-1}b^1, b^2$ are linearly independent. Choose $v^{j+2},\dots,v^m$ so that $b^1,\dots,A^{j-1}b^1,b^2,v^{j+2},\dots,v^m$ is a basis of $\mathbf{R}^m$. One could form the basis from b^i's and A^kb^k's, but this is not necessary. Make the change of coordinates $x = P\bar{x}$ where $P = (b^1,\dots,A^{j-1}b^1,b^2,v^{j+1},\dots,v^m)$. Then

$$\bar{x}' = \bar{A}x + \delta^1u_1 + \delta^{j+1}u_2 + P^{-1}b_3u_3 + \cdots + P^{-1}b_ru_r,$$

where

$$\bar{A} = \begin{pmatrix} A_{11} & A_{12} \\ 0 & A_{22} \end{pmatrix}$$

and

$$A_{11} = \begin{pmatrix} 0 & \cdot & \cdot & \cdot & 0 & a_0 & \alpha_0 \\ 1 & \ddots & & & & a_1 & \alpha_1 \\ 0 & \ddots & \ddots & & & \vdots & \vdots \\ \vdots & & \ddots & \ddots & 0 & \vdots & \vdots \\ \vdots & & & \ddots & 1 & a_{j-1} & \alpha_{j-1} \\ 0 & \cdot & \cdot & \cdot & 0 & 0 & \alpha_j \end{pmatrix}.$$

Now add linear feedback $\delta^{j+1}c\bar{x}_j$, $c \neq 0$. Then

$$\bar{x}' = \bar{A}_1\bar{x} + \bar{B}u,$$

where

$$\bar{A}_1 = \bar{A} + \delta^{j+1}c = \begin{pmatrix} \hat{A}_{11} & A_{12} \\ 0 & A_{22} \end{pmatrix}$$

and

$$\hat{A}_{11} = \begin{pmatrix} 0 & & & & a_0 & \alpha_0 \\ 1 & \ddots & & & \cdot & \cdot \\ 0 & \ddots & \ddots & & \cdot & \cdot \\ \vdots & & \ddots & \ddots & \cdot & \cdot \\ & & \cdot & 0 & a_{j-1} & \\ 0 & \dots & 0 & 1 & c & \alpha_j \end{pmatrix}$$

Hence

$$\delta^1,\ \bar{A}_1\delta^1 = \delta^2,\dots,\bar{A}_1^{j-1}\delta^1 = \delta^j,\ \bar{A}_1\delta^1 = \begin{pmatrix} a_0 \\ a_1 \\ \vdots \\ a_{j-1} \\ c \\ 0 \\ \vdots \\ 0 \end{pmatrix}$$

are linearly independent for $c \neq 0$. In the original coordinates this corresponds to $b^1, A_1b^1,\dots,A_1^jb^1$ linearly independent; $A_1 = A+BC_1$ and $\{A_1,B\}$ is controllable (Proposition 15.11). If $j+1 = m$, stop. Since the same algorithm applies to any nonzero column of B, this completes the proof. □

16.4. Theorem. A pair $\{A,B\}$ is σ-controllable if and only if $\{A,B\}$ is controllable.

Proof: The sufficiency follows from the above proposition and Proposition 10.4. To prove the necessity assume $\{A,B\}$ has an assignable spectrum. Select C_0 so that $(A+BC_0)^n \to 0$ as $n \to \infty$ $(r(A+BC_0) < 1)$, and C_1 such that

$\sigma(A+BC_1) = \{e^{\frac{2\pi k}{m} i}; k = 0,\dots,m-1\}$, the $m^{\underline{th}}$-roots of unity. Then $(A+BC_1)^m = I$. Suppose $v^T A^n B = 0$ for all $n \geq 0$. Then, for any C,

$$v^T(A+BC)^n = v^T(A+BC)(A+BC)^{n-1} = v^T A(A+BC)^{n-1}$$

$$= \cdots = v^T A^n \quad \text{for all } n.$$

Hence

$$v^T[(A+BC_0)^n - (A+BC_1)^n] = 0$$

and

$$v^T[(A+BC_0)^{km} - I] = 0, \quad \text{all } k \geq 0.$$

Letting $k \to \infty$, implies $v^T = 0$, and, by Theorem 15.8, $\{A,B\}$ is controllable. □

Hence, a sufficient condition for being able to stabilize $x' = Ax + Bu$ by linear feedback control is controllability. We will say that the system $x' = Ax + Bu$, or the pair $\{A,B\}$, is <u>stabilizable</u> if there is a matrix C such that $A+BC$ is stable. What we can expect when a system is not controllable, is that a part of the system will be controllable and a part not controllable. The system will be stablizable if and only if the uncontrollable part is asymptotically stable.

Let us see how one goes about decomposing a system into its controllable and uncontrollable parts. We will look at the case $m = 3$ and $r = 2$, and this is enough to show how to do the general case. Assume that the system

$$x' = Ax + Bu = Ax + b^1 u_1 + b^2 u_2$$

is not completely controllable. There are essentially only

two cases. The case of no control $(b_1 = b_2 = 0)$ is trivial.

Case 1. b_1 and Ab_1 linearly independent, $A^2b^1 = a_0b^1 + a_1Ab^1$, $b^2 = \beta_0 b^1 + \beta_1 Ab^1$. Taking $P = (b^1, Ab^1, v^3)$, P nonsingular, and $x = P\bar{x}$, we obtain $(Pv^3 = \alpha_0 b^1 + \alpha_1 Ab^1 + \alpha_3 v^3)$

$$\bar{x}_1' = a_0\bar{x}_2 + \alpha_0\bar{x}_3 + u_1 + \beta_0 u_2$$

$$\bar{x}_2' = \bar{x}_1 \quad + a_1\bar{x}_2 + \alpha_1\bar{x}_3 + \beta_1 u_2$$

$$\bar{x}_3' = \alpha_3\bar{x}_3.$$

The controllable variables are $\bar{x}_1$, $\bar{x}_2$ and $\bar{x}_3$ is uncontrollable. The system is stabilizable if and only if $|\alpha_3| < 1$; i.e., if and only if the uncontrollable part is already asymptotically stable.

Case 2. $b^1 \neq 0$, $Ab^1 = a_0 b^1$, $b^2 = \beta_0 b^1$ (there is really only one control variable). Select v^2, v^3 so that b^1, v^2, v^3 is a basis. Let $x = P\bar{x}$, $P = (b^1, v^2, v^3)$, $Av^2 = \alpha_{12}b^1 + a_{22}v^2 + \alpha_{32}v^3$, $Av^3 = \alpha_{13}b^1 + \alpha_{23}v^2 + \alpha_{33}v^3$.

$$\bar{x}' = \begin{pmatrix} a_0 & \alpha_{12} & \alpha_{13} \\ 0 & \alpha_{22} & \alpha_{23} \\ 0 & \alpha_{32} & \alpha_{33} \end{pmatrix}\bar{x} + \begin{pmatrix} 1 & \beta_0 \\ 0 & 0 \\ 0 & 0 \end{pmatrix}u,$$

hence we see the system is stabilizable if and only if

$A_{22} = \begin{pmatrix} \alpha_{22} & \alpha_{23} \\ \alpha_{32} & \alpha_{33} \end{pmatrix}$ is stable; $\bar{x}^1$ is controllable, $\bar{x}^2$ and $\bar{x}^3$ are not, and A_{22} is the matrix of the uncontrollable part.

In the general case we can select a basis for the controllable part of $x' = Ax + Bu$, complete that basis to obtain a basis of $\mathbf{R}^m$. Take this basis as a basis of new coordinates (new state variables), and obtain

$$\bar{x}' = \bar{A}\,\bar{x} + \bar{B}u, \qquad \text{where}$$

$$\bar{A} = \begin{bmatrix} A_{11} & A_{12} \\ 0 & A_{22} \end{bmatrix}, \quad \bar{B} = \begin{bmatrix} B_1 \\ 0 \end{bmatrix}.$$

with $\bar{x} = \binom{\xi}{\eta}$, where ξ is the controllable state vector and η is the uncontrollable state vector. Then

$$\begin{aligned} \xi' &= A_{11}\xi + A_{12}\eta + B_1 u \\ \eta' &= A_{22}\eta. \end{aligned}$$

The pair (A_{11}, B_1) is controllable and A_{22} is the matrix of the uncontrollable part. <u>The system</u> $x' = Ax + Bu$ <u>is stabilizable if and only if</u> A_{22} <u>is stable</u>. The dimension of the uncontrollable part is $m\text{-rank}(Y(m))$.

16.5. <u>Exercise</u>. Determine the matrices B for which the system

$$2x' = Ax + Bu, \qquad A = \begin{bmatrix} 2 & -2 & 4 \\ 0 & 1 & 2 \\ 1 & -1 & 2 \end{bmatrix},$$

is (a) controllable? (b) stabilizable?

16.6. <u>Exercise</u>. a. Describe an algorithm for determining the decomposition of $x' = Ax + Bu$ into its controllable and uncontrollable parts when $m = 4$.

b. Describe an algorithm in the general case.

17. Minimum energy control. Minimum time-energy feedback controls.

The positive semidefinite symmetrix matrix

$$W(n) = Y(n)Y^T(n) = \sum_{j=0}^{n-1} A^j BB^T (A^j)^T \tag{17.1}$$

introduced in Section 15 plays an important role in the theory of linear control systems and is called the controllability grammian (see Nering [1], p. 150 for the definition of the grammian of a set of vectors; $W(n)$ is the grammian of the matrices $B, AB, \ldots, A^{n-1}B$).

For n fixed and $v \in \mathbf{R}^m$ consider the control defined by

$$v^*(n-j-1) = (A^j B)^T v, \quad j = 0,\ldots,n-1. \tag{17.2}$$

For this control

$$\pi(n,v^*,x) = A^n x + W(n)v,$$

and, if there is a v satisfying

$$W(n)v = y - A^n x, \tag{17.3}$$

then the control v^* brings the systems from x to y in time n.

The theorem we will now prove includes the converse of what we have just shown. For $x, y \in \mathbf{R}^m$ the inner product is $(x,y) = x^T y$. Since $W(n)$ is symmetric, we know for each $w \in \mathbf{R}^m$ that $w = w^1 + w^0$ where $w^1 \in$ image $W(n)$, $W(n)w^0 = 0$, and $(w^0,w^1) = 0$ -- the image and the kernel of $W(n)$ are orthogonal and $\mathbf{R}^m$ is the direct sum of image$W(n)$ and ker $W(n)$.

17.1. <u>Proposition</u>. Image $W(n)$ = image $Y(n)$.

<u>Proof</u>: We saw above that image $W(n) \subset$ image $Y(n)$. (this is also an immediate consequence of (17.1).) It remains to show that image $Y(n) \subset$ image $W(n)$. If $w \in$ image $Y(n)$, there is a control u for which $w = \sum_{j=0}^{n-1} A^j Bu(n-j-1)$. By the remark just above this proposition $w = w^1 + w^0$, where $w \in$ image $W(n)$, $W(n)w^0 = 0$ and $(w^0,w) = ||w^0||^2$. From (17.1), we see that $W(n)w^0 = 0$ implies $w^{0^T} A^j B = 0$, $j = 0,1,\ldots,n-1$. Now

$$(w^0,w) = ||w^0||^2 = \sum_{j=0}^{n-1} w^{0^T} A^j Bu(n-j-1) = 0,$$

and $w^0 = 0$; i.e., $w \in$ image $W(n)$ and image $Y(n) \subset$ image $W(n)$. Hence image $W(n)$ = image $Y(n)$. □

From this proposition and Proposition (5) we know that, if there is a control u such that $y = \pi(n,u,x)$, then $w = y - A^n x \in$ image $W(n)$. Therefore, with any v such that $W(n)v = w$, we have that $y = \pi(n,v^*,x^0)$, where $v^*(j)$ is defined by (17.2) for $j = 0,\ldots,n-1$. We want to show that this $v^*(j)$ is an optimal control -- it is, as we shall explain, the minimum "energy" control. We now have that

$$\sum_{j=0}^{n-1} (v,A^j Bu(n-j-1)) = \sum_{j=0}^{n-1} (v,A^j Bv^*(n-j-1)$$

or

$$\sum_{j=0}^{n-1} ((A^j B)^T v, u(n-j-1))$$

$$= \sum_{j=0}^{n-1} (v^*(j),u(j)) = \sum_{j=0}^{n-1} ||v^*(j)||^2.$$

Hence

$$\sum_{j=0}^{n-1} ||u(j)||^2 = \sum_{j=0}^{n-1} [||v^*(j)||^2 + ||u(j)-v^*(j)||^2]$$

and, if $u(j) \neq v^*(j)$ for some $j = 0,1,\ldots,n-1$, then

$$\sum_{j=0}^{n-1} ||u(j)||^2 > \sum_{j=0}^{n-1} ||v^*(j)||^2.$$

We have proved that

17.2. Theorem. If y is reachable from x in time n, then the control v^* defined by (17.2) minimizes the control energy $E(u) = \sum_{j=0}^{n-1} ||u(j)||^2$; in fact, if u is any other control that does this and $u(j) \neq v^*(j)$ for some $j = 0,1,\ldots,n-1$, then

$$E(u) > E(v^*) = (v, W(n)v).$$

17.3. Exercise. Another measure of the cost of control is

$$C(u) = \sum_{j=0}^{n-1} (u(j), Qu(j)),$$

where Q is an $r \times r$ positive definite symmetric matrix. There is then a nonsingular matrix P such that $Q = P^T P$. Apply Theorem 17.2 to the system $x' = Ax + \hat{B}\hat{u}$, where $\hat{B} = BP^{-1}$ and $\hat{u} = Pu$, to obtain the control that minimizes $C(u)$.

We know that if y is reachable from x in time n, that $y - A^n x \in$ image $W(n)$ and there is a unique minimum energy control

$$\varepsilon(j; x,y,n) = v^*(j) = (A^{n-j-1}B)^T v, \quad j = 0,\ldots,n-1, \tag{17.4}$$

where v is any solution of

$$W(n)v = y - A^n x. \tag{17.5}$$

initial state and does this with minimum energy. Of all controls $v: \mathbf{R}^m \to \mathbf{R}^r$, that bring each initial state to y in

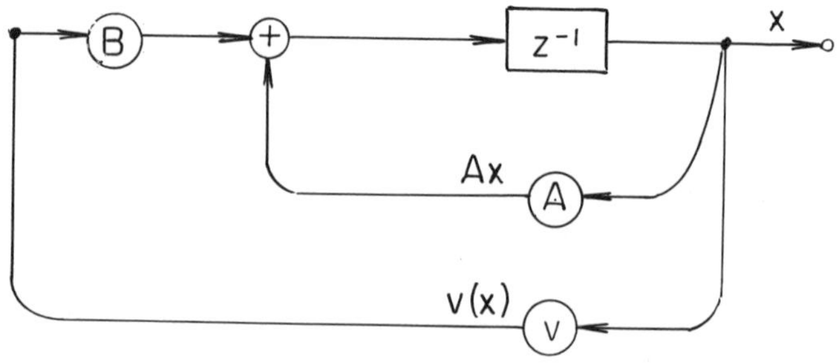

Feedback Control

Figure 17.1

finite time, v^0 does it in the least time, and, of all controls that do this in least time, v^0 does it with the least energy (minimizes $\sum_{n=0}^{n(x^0)} ||v(x)||^2$ along trajectories). We will call this unique optimal control v^0 the <u>minimal time-energy control</u>. Here we assume no constraints on the allowable control laws $v(x)$.

For $x \in \mathbf{R}^m$, $x \neq y$, define

$$v^0(x) = \varepsilon(0,x,y,n(x)) = (A^{n(x)-1}B)^T v, \tag{17.8}$$

where $\hat{W}(x) = W(n(x))$ and

$$\hat{W}(x)v = y - A^{n(x)}x. \tag{17.9}$$

Let $v: \mathbf{R}^m \to \mathbf{R}^r$ be any control law. Relative to the system (17.7), $\hat{\pi}(n,v,x^0)$ denotes the solution of (17.7) satisfying $\hat{\pi}(0,v,x^0) = x^0$. Then $u(n) = v(\hat{\pi}(n,v,x^0))$ is an open loop control for the system $x' = Ax + Bu$ and $\pi(n,u,x^0) = \hat{\pi}(n,v,x^0)$ for all $n \geq 0$.

The uniqueness of this control follows from Theorem 17.3. It also can be seen directly since, if v^1 and v^2 are solutions of (17.5), $W(n)(v^1-v^2) = 0$ and from (17.1) we see that this implies $(A^{n-j-1}B)^T v^1 = (A^{n-j-1}B)^T v^2$ for $j = 0,\ldots,n-1$. Let

$$x^1 = \pi(1;\ v^*,x) = Ax + Bv^*(0).$$

Then the minimum energy control to bring x^1 to y in time $n-1$ is given by

$$\varepsilon(j;\ x^1,y,n-1) = \varepsilon(j+1;x,y,n), \quad j = 0,\ldots,n-2. \qquad (17.6)$$

This is an example of the principle of optimality -- each portion of an optimal trajectory is optimal. Otherwise, this would contradict Theorem 17.2. If

$$x^i = \pi(i,v^*,x),$$

we have

$$\varepsilon(j;\ x^i,y,n-i) = \varepsilon(j+i,x,y,n), \quad j = 0,\ldots,n-i-1.$$

Let us assume now that the pair $\{A,B\}$ is controllable. Then for each $x \in \mathbb{R}^m$ -- each initial state -- there is a unique integer $n(x) \leq m$ such that $y - A^{n(x)} \in$ image $Y(n(x)) =$ image $W(n(x))$ and $y - A^{n(x)-1}x \notin W(n(x)-1)$. The time $n(x)$ is the <u>minimum time</u> to reach y from x.

We want to show now that, if $\{A,B\}$ is controllable, we can synthesize a feedback control law $v^0(x)$ with the property that for the (closed loop) control system (Figure 17.1)

$$x' = Ax + Bv(x) \qquad (17.7)$$

$v^0(x)$ brings the system to y in minimum time from each

17.4. <u>Corollary</u>. If the pair $\{A,B\}$ is controllable, then the control v^0 defined by (17.8) and (17.9) is the unique minimal time-energy control for (17.7). The minimum time to reach y from x is $n(x)$ and the minimum energy is

$$E(x) = \sum_{j=0}^{n(x)-1} ||v^0(\hat{\pi}(j,v^0,x)||^2.$$

<u>Proof</u>: By what has been pointed out above it is easy to see that

$$v^0(\hat{\pi}(j,v^0,x)) = \varepsilon(j,x,y,n(x)), \quad j = 0,1,\ldots,n(x)-1$$

and the conclusion follows from Theorem 17.2. □

It is important to note in the above that $v^0(y)$ has not been defined; $\hat{\pi}(n(x),v^0,x) = y$ and $n(y) = 0$. If y is an equilibrium point of the uncontrolled system $(Ay = y)$, then defining $v^0(y) = 0$ we see that y is an equilibrium point of the controlled system (17.7), and the <u>minimal time-energy control</u> v^0 <u>has made</u> y <u>a global attractor</u>. If y is not an equilibrium state of the uncontrolled system and the equation

$$Bv = (I-A)y$$

has a solution, then defining $v^0(y)$ to be any solution makes y an equilibrium point of the controlled system (17.7) and makes y a global attractor. This is another important consequence of controllability.

17.5. <u>Corollary</u>. Assume that $\{A,B\}$ is controllable and that $(I-A)y$ is in the image of B. Define $v^0(y)$ to be any solution of $Bv = (I-A)y$. Then the minimal <u>time-energy control</u> v^0 makes y an equilibrium point of (17.7) and a finite time global attractor; in fact, $\hat{\pi}(n,v^0,x) = y$ for all $n \geq m$ and all $x \in \mathbf{R}^m$.

If $Bv = (I-A)y$, we can always arbitrarily define $v^0(y) = u^0 \in \mathbf{R}^m$ ($u^0 = 0$, for example). Then

$$\pi(n(x)+1,v^0,x) = Ay + Bu^0 = x^1.$$

The motion from x^1 will then return again to y in time $n(x^1)$, and this is a periodic motion through y of period $n(x^1) + 1$. If the objective was simply to reach y, this is no problem. If the process is to continue, then one would have to worry about the choice of $v^0(y)$. The best that one could then do is to select the $v^0(y)$ that makes the oscillation through y as "small" as possible. Is, perhaps, a business cycle this type of phenomenon? <u>Removing the control when the desired state is reached could be the worst thing to do</u>, and this is an aspect of control, if neglected, in the real world where there are delayed effects can have serious consequences.

Let us look briefly at another practical problem that we will not attempt to analyze although we will suggest what might be a good solution. We continue to assume that $\{A,B\}$ is completely controllable. In the definition of the time-energy optimal control v^0 (equations (17.8) and (17.9)) it was only necessary to say that v is any solution of

$$\hat{W}(x)v = y - A^{n(x)}x.$$

However, in the real world the system is subject to perturbations (errors in state estimation in determining parameters for the model, etc.,), and if this were to be automated (computer controlled) it might make a difference which solution is selected. One way to select each time a unique solution is to use one of the generalized inverses of $\hat{W}(x)$, and intuitively at least a good choice would seem to be the Moore-Penrose inverse (for a discussion of the existence, uniqueness, computation, and general properties of generalized inverses see Ben-Israel and Greville [1]). We will state only some basic facts.

Let W be any real $m \times m$ matrix. The Moore-Penrose inverse W^+ of W is the unique matrix with the following properties:

$$WW^+W = W$$
$$W^+WW^+ = W^+$$
$$(W^+W)^\tau = W^+W$$
$$(WW^+)^\tau = WW^+.$$

There always exists a unique solution W^+, and there are algorithms (iterate methods) for computing successive approximations. Since W^+ is unique, $W^+ = W^{-1}$ if W is nonsingular. Relative to the equation

$$Wx = b$$

W^+b is the "best" approximate solution in the following sense:

17.6. <u>Proposition</u>. Given $b \in R^m$, $x^+ = A^+b$ has the property that for all $x \neq x^+$

$$||Wx - b|| > ||Wx^+ - b||$$

or

$$||Wx - b|| = ||Wx^+ - b|| \quad \text{and} \quad ||x|| > ||x^+||.$$

Using $\hat{W}^+(x)$ gives as a formula for the minimal time-energy control

$$v^0(x) = (A^{n(x)-1}B)^T W^+(n(x))(y - A^{n(x)}x). \tag{17.10}$$

Because of perturbations and approximations (17.9) may have no solutions but (17.10) always defines a control. One can modify this definition to define a control law even where y cannot be reached from x but the analysis is more complicated.

18. Observability. Observers. State Estimation. Stabilization by dynamic feedback.

Up to now in our consideration of the linear system (Figure 18.1)

$$x' = Ax + Bu \tag{18.1}$$

we have assumed that the output is the state of the system x.

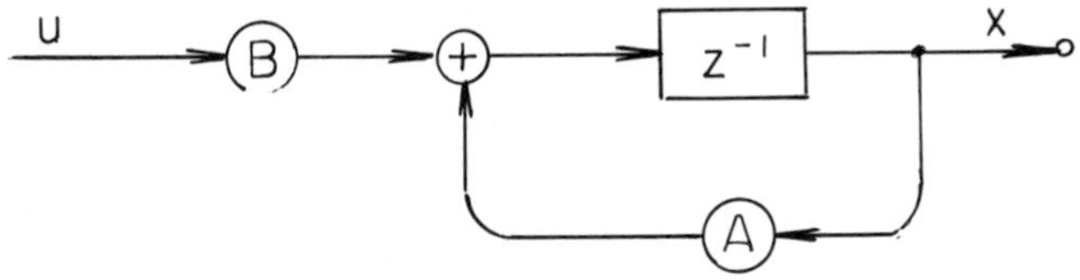

Figure 18.1

In actuality it may not always be possible to observe the state of the system directly, and we want now to add another aspect of reality by considering the linear input-output system

$$S: \quad \begin{aligned} x' &= Ax + Bu \\ y &= Cx + Du, \end{aligned} \tag{18.2}$$

where u is the input and y is the output (Figure 18.2). The matrices are all real; A is $m \times m$, B is $m \times r$, C is $s \times m$ and D is $s \times r$. Given an input $u: J_0 \to \mathbf{R}^r$ and an

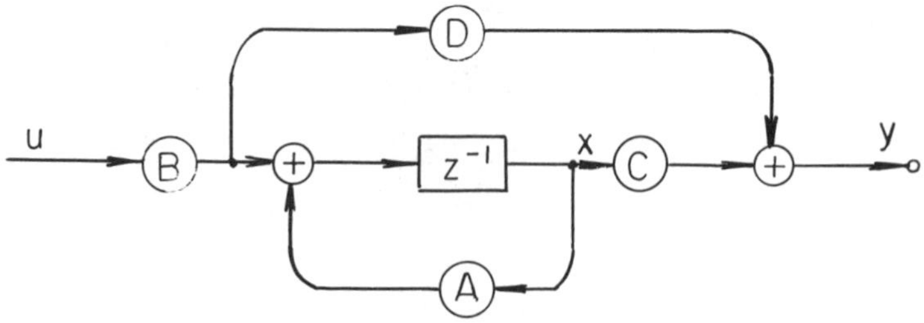

The System S

Figure 18.2

initial state $x^0 \in \mathbf{R}^m$, $\pi(n,x^0,u)$ is the solution of (18.1) satisfying $\pi(0,x^0,u) = x^0$. The output at time n is then

$$\eta(n,x^0,u) = C\pi(n,x^0,u) + Du(n).$$

The question of observability of the initial state is the following: The system starts at time $n = 0$ in some initial state x^0, and the output $y(n) = \eta(n,x^0,u)$ is observed for $n = 0,1,\ldots,n_1$. Do these observations of the output make it possible to determine uniquely the initial state x^0? In other words, is the initial state observable? Because we are limiting ourselves to linear input-output systems this question is greatly simplified. By linearity we know that

$$\begin{aligned}\pi(n,x^0,u) &= \pi(n,x^0,0) + \pi(n,0,u) \\ &= A^n x^0 + \pi(n,0,u),\end{aligned}$$

and

$$\begin{aligned}\eta(n,x^0,u) &= \eta(n,x^0,0) + \eta(n,0,u) \\ &= CA^n x^0 + \eta(n,0,u).\end{aligned} \tag{18.3}$$

The output $\eta(n,x^0,u)$ is the zero input response plus the zero initial state output. An initial state x^0 is said to be observable if for some $n_1 \geq 0$ and some input u, $\eta(n,x^0,u) = \eta(n,x^1,u)$ for $n = 0,1,\ldots,n_1$ implies $x^1 = x^0$. If each $x^0 \in R^m$ is observable we say that the system (18.2) is observable. (Sometimes the additional adjective "completely" is used.) From (18.3) we see that B, D, and u play no role, and we can always take $u = 0$. The following theorem shows that there is a dual relationship between observability and controllability, and all of our results on controllability are applicable. We give two proofs. The first shows that duality follows immediately from Theorem 15.8, and the second gives an explicit expression for x^0 when (18.2) is observable.

18.1. Theorem. The input-output system (18.2) is observable if and only if the pair $\{A^T, C^T\}$ is controllable.

Proof:1. Take $u = 0$, and consider the outputs

$$\begin{aligned} y(0) &= Cx^0 \\ y(1) &= CAx^0 \\ &\vdots \\ y(m-1) &= CA^{m-1}x^0. \end{aligned} \tag{18.4}$$

Define

$$\hat{y}(m) = \begin{pmatrix} y(0) \\ y(1) \\ \vdots \\ y^{(m-1)} \end{pmatrix} \quad \text{and} \quad Y(m) = \begin{pmatrix} C \\ CA \\ \vdots \\ CA^{m-1} \end{pmatrix}.$$

The system of equations (18.4) can then be written

$$\hat{y}(m) = \hat{Y}(m)x^0;$$

$\hat{Y}(m)$ is an $ms \times m$ matrix and is a linear transformation of $\mathbf{R}^m$ into $\mathbf{R}^{ms}$. By Theorem 15.8 controllability of $\{A^T, C^T\}$ is equivalent to $\hat{Y}(m)$ is one-to-one (Rank $\hat{Y}(m) = m$), and this is equivalent to observability of (18.2). □

Proof 2. Necessity. Assume $\{A^T, C^T\}$ is not controllable. By Theorem 15.8 this implies there is an $x^0 \neq 0$ such that $CA^n x^0 = 0$ for all $n \in J_0$, and clearly (18.3) is not observable -- x^0 cannot be distinguished from the zero state.

Sufficiency. Assume $\{A^T, C^T\}$ is controllable. By Theorem 15.8 this implies $\hat{W}(m) = \hat{Y}^T(m)\hat{Y}(m)$ is positive definite. Then $\hat{Y}^T(m)\hat{y}(m) = \hat{W}(m)x^0$, and $x^0 = \hat{W}^{-1}(m)\hat{Y}^T(m)\hat{y}(m)$; x^0 is uniquely determined by $\hat{y}(m)$. □

The observability of (18.2) depends only upon the pair $\{A,C\}$, and when we say $\{A,C\}$ is observable, we mean s is observable. We then have from Theorems 15.8 and 16.4 the following corollary.

18.2. Corollary. The following are equivalent:

(i) The system (18.2) (the pair $\{A,C\}$) is observable.

(ii) rank $\hat{Y}(m) = m$.

(iii) $\hat{Y}(n)x = 0$ for all $n \geq 0$ implies $x = 0$.

(iv) $\hat{W}(m)$ is positive definite.

(v) Given $\sigma_0 = \{\lambda_1, \lambda_2, \ldots, \lambda_m\}$ with $\sigma_0 = \overline{\sigma}_0 = \{\overline{\lambda}_1, \overline{\lambda}_2, \ldots, \overline{\lambda}_m\}$ there is a matrix R such that $\sigma(A+RC) = \sigma_0$.

18.3. Example. Consider the input-output system shown in Figure 18.3. The equations for the system are

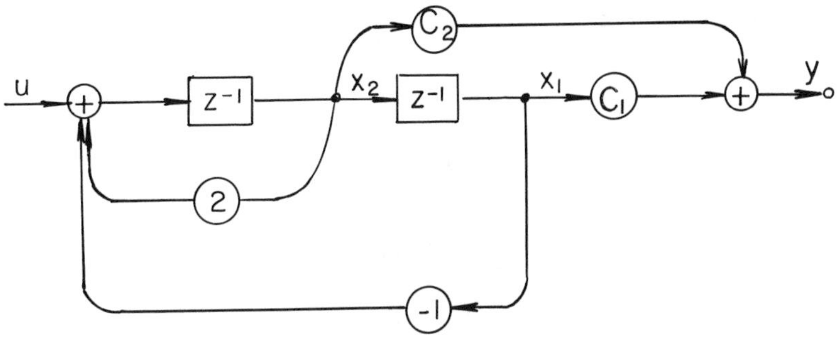

Figure 18.3

$$S: \quad \begin{aligned} x_1' &= x_2 \\ x_2' &= -x_1 + 2x_2 + u \\ y &= c_1x_1 + c_2x_2; \end{aligned}$$

$A = \begin{bmatrix} 0 & 1 \\ -1 & 2 \end{bmatrix}$, $B = \begin{bmatrix} 0 \\ 1 \end{bmatrix}$, $C = (c_1, c_2)$, and $D = 0$. A second order equation for the system is $\theta'' - 2\theta' + \theta = u$, where $\theta = x_1$. Since $CA = (-c_2, c_1+2c_2)$, the system is observable if and only if $c_1 + c_2 \neq 0$. In this case

$\hat{Y}(2) = \begin{bmatrix} c_1 & c_2 \\ -c_2 & c_1+2c_2 \end{bmatrix}$ and, if $c_1 + c_2 \neq 0$, $\hat{Y}(m)$ is non-singular. Note that when $c_1 + c_2 = 0$ $y = c_2(x_2 - x_1) = c_2(\theta' - \theta)$ and $y' - y = c_2u$.

Consider the particular case $c_1 = 1$, $c_2 = 0$. The system is then both controllable and observable, and let us see if through direct linear feedback it can be stabilized (Figure 18.4). The equations are then $(u = \alpha y = \alpha x_1)$

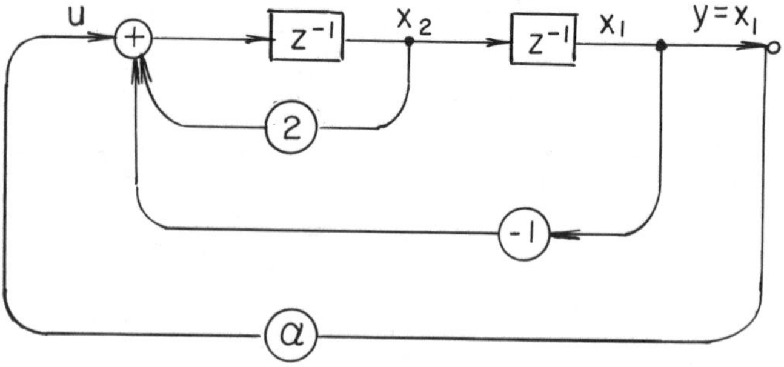

Figure 18.4

$$x_1' = x_2$$
$$x_2' = (\alpha-1)x_1 + 2x_2$$
$$y = x_1.$$

The characteristic polynomial is $\lambda^2 - 2\lambda + (1-\alpha)$. The conditions for asymptotic stability (see Example 5.4) are $|1-\alpha| < 1$ and $2 < 2-\alpha$, and we see that the system can never be stabilized (made asymptotically stable) with this type of direct feedback. Note, however, the feedback scheme shown in Figure 18.5. This is an example of indirect or dynamic feedback. The control is generated by a difference equation.

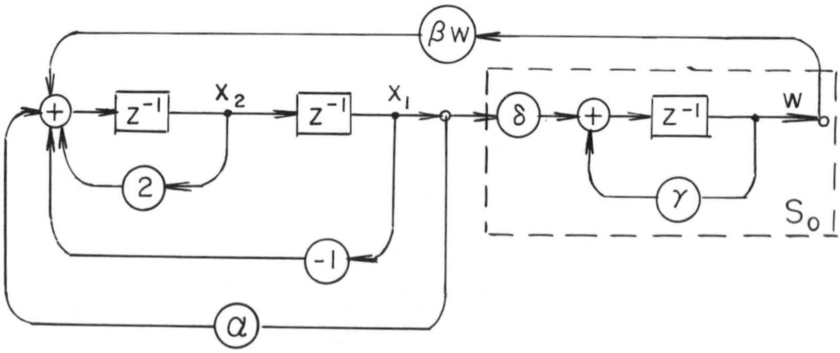

Figure 18.5

The equations for this feedback scheme are

$$\begin{aligned} & x_1' = x_2 \\ & x_2' = (\alpha-1)x_1 + 2x_2 + \beta w \\ S_0: \quad & w' = \gamma w + \delta x_1 \quad . \end{aligned} \tag{18.5}$$

The characteristic polynomial is

$$\phi(\lambda) = \lambda^3 - (2+\gamma)\lambda^2 + (2\gamma-\alpha+1)\lambda + \gamma(\alpha-1) - \beta\delta.$$

Clearly by the choice of γ, α, and $\beta\delta$ we can make the eigenvalues anything we please, and thus can stabilize the system.

Let us continue with this example a bit more and see if we can understand just what it is that the device S_0 does. The role of S_0 becomes clear if we introduce the new variable

$$\varepsilon = w - rx_1 - x_2,$$

where r is for the moment arbitrary. The equations for (18.5) then become

$$\begin{aligned} x_1' &= x_2 \\ x_2' &= (\alpha+r\beta-1)x_1 + (2+\beta)x_2 + \beta\varepsilon \\ \varepsilon' &= (\gamma-\beta)\varepsilon + (r\gamma-r\beta+\delta-\alpha+1)+(\gamma-\beta-r-2)x_2, \end{aligned}$$

and, selecting

$$\begin{aligned} r &= \gamma - \beta - 2 \\ \alpha &= r(\gamma-\beta) + \delta + 1, \end{aligned}$$

we have

$$\begin{aligned} x_1' &= x_2 \\ x_2' &= (\alpha+r\beta-1) + (2+\beta)x_2 + \beta\varepsilon \\ \varepsilon' &= (\gamma-\beta)\varepsilon; \end{aligned}$$

i.e.

$$\begin{pmatrix} x_1' \\ x_2' \\ \varepsilon' \end{pmatrix} = \hat{A} \begin{pmatrix} x_1 \\ x_2 \\ \varepsilon \end{pmatrix},$$

where

$$\hat{A} = \begin{bmatrix} 0 & 1 & 0 \\ \alpha + r\beta - 1 & 2+\beta & \beta \\ 0 & 0 & \gamma - \beta \end{bmatrix}$$

Now

$$\phi(\lambda) = \det(\lambda I - \hat{A}) = (\lambda - (\gamma-\beta))(\lambda^2 - (2+\beta)\lambda + 1 - \alpha - r\beta),$$

and since we are still free to choose γ, β and δ, we can arbitrarily assign the eigenvalues and can thereby stabilize the system. Note that this requires $|\gamma-\beta| < 1$. Since $\varepsilon(n) = (\gamma-\beta)^n \varepsilon(0)$, we see that $\varepsilon(n) \to 0$ as $n \to \infty$;

$$w(n) - rx_1(n) - x_2(n) = (\gamma-\beta)^n (w(0) - rx_1(0) - x_2(0)) \to 0 \text{ as } n \to \infty.$$

<u>The output</u> $w(n)$ <u>of</u> S_0 <u>gives an asymptotic estimate of</u> $rx_1(n) + x_2(n)$. A device (algorithm) such as S_0 is called an "observer" or "asymptotic state estimator". Since the output of the original system is x_1, $w - rx_1$ estimates x_2, and hence at each time n we know $x_1(n)$ and have an estimate of $x_2(n)$. The system is controllable, and linear feedback of the state estimate makes it possible to stabilize the system. For instance, selecting $\beta = \gamma = -2$, $r = -2$, $\delta = -4$ and $\alpha = -3$ makes $\psi(\lambda) = \lambda^3$ and $\hat{A}^3 = 0$; in fact,

$$\hat{A} = \begin{pmatrix} 0 & 1 & 0 \\ 0 & 0 & -2 \\ 0 & 0 & 0 \end{pmatrix}, \quad \hat{A}^2 = \begin{pmatrix} 0 & 0 & -2 \\ 0 & 0 & 0 \\ 0 & 0 & 0 \end{pmatrix},$$

$$\begin{pmatrix} x_1(1) \\ x_2(1) \\ \varepsilon(1) \end{pmatrix} = \begin{pmatrix} x_2(0) \\ -2\varepsilon(0) \\ 0 \end{pmatrix}, \quad \begin{pmatrix} x_1(2) \\ x_2(2) \\ \varepsilon(2) \end{pmatrix} = \begin{pmatrix} -2\varepsilon(0) \\ 0 \\ 0 \end{pmatrix}$$

and $\begin{pmatrix} x_1(n) \\ x_2(n) \\ \varepsilon(n) \end{pmatrix} = 0$ for $n \geq 3$.

Note that $w(n) = 2x_1(n) - x_2(n)$ for all $n \geq 1$, so that in unit time the output of the original system has been in theory estimated exactly.

18.4. Exercise. Select for the system (18.5) γ, α and δ so that its characteristic polynomial is λ^3 (all eigenvalues are zero). Show in this case that S_0 identifies $-\frac{2}{\beta}(-2x_1 + x_2)$.

We now want to look at the general case of an input-output system S and an observer S_0 (see Figure 18.6).

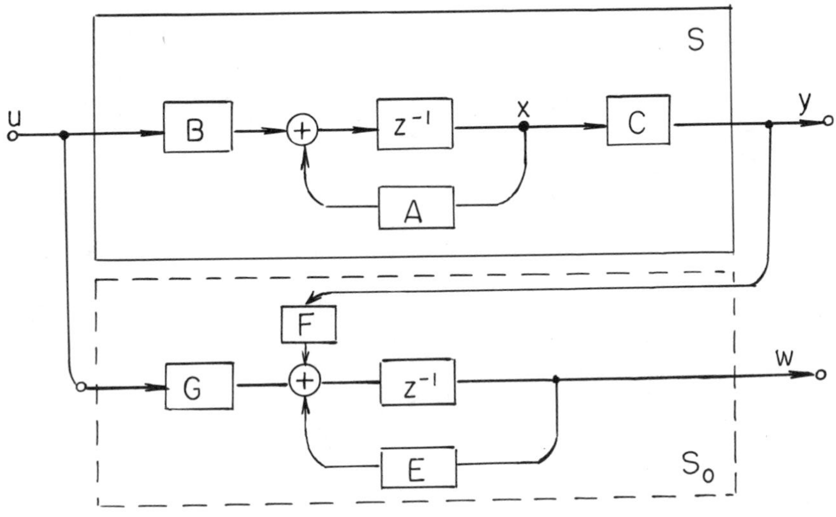

Figure 18.6

The equations for S and S_0 are

$$S: \quad \begin{aligned} x' &= Ax + Bu \\ y &= Cs \end{aligned}$$
$$S_0: \quad w' = Ew + Fy + Gu. \tag{18.6}$$

This is the same input-output system as before (equation 18.2) except that we have taken $D = 0$. Nothing is lost by duing this since Du can always be subtracted from the output y. Observability of S has to do with the possibility of determining the state $x(n)$ from a knowledge of the outputs $y(n), y(n+1),\dots,y(n+m-1)$. If S is observable, we know that

$$x(n) = \hat{W}^{-1}(m)\hat{K}^T(m)\hat{y}(n+m);$$

the state at time n is determined at time $n+m-1$. The problem we want to consider now is the determination of $x(n)$ from the observation of the output $y(0), y(1),\dots,y(n)$; we would like to know or have a good estimate, of the state $x(n)$ at time n (<u>no time delay</u>). Let us consider first the problem of having S_0 estimate Kx, where K is a given $s_0 \times m$ matrix; s_0 will be the dimension of the observer S_0. The output $w(n)$ of the observer depends upon the initial estimate $w(0)$ of $Kx(0)$ and the choice of E, F, and G. Let $\varepsilon(n) = w(n) - Kx(n)$, the error in the estimate. A simple computation gives

$$\varepsilon' = E\varepsilon + (EK+FC-KA)x + (G-KB)u. \tag{18.7}$$

Hence, if

$$G = KB$$

and

$$EK + FC = KA, \tag{18.8}$$

$$\varepsilon' = E\varepsilon \quad \text{and} \quad w(n) - Kx(n) = E^n(w(0)-Kx(0)).$$

Thus, if we could find an F and a stable E satisfying (18.8) we would have

$$w(n) - Kx(n) \to 0 \quad \text{as} \quad n \to \infty,$$

and in this case we will say that S_0 observes Kx or S_0 is an observer of Kx. How fast $w(n)$ converges to $x(n)$ depends upon the eigenvalues of E and the initial estimate $w(0)$ of $Kx(0)$. Now C, K, and A are given, and it may or may not be possible to find an F and a stable E satisfying (18.8). For instance, if we want S_0 to observe x, then $K = I$ and equation (18.8) becomes

$$E = A - FC.$$

This is not an unfamiliar equation. We know by Corollary 18.2 that, if S is observable, that by the choice of F we can arbitrarily assign the eigenvalues of E. We have therefore the following result.

18.5. Proposition. Given an observable system S it is always possible to construct an observer S_0 of the state of S with E having arbitrarily assigned eigenvalues.

18.6. Example. Let us now construct a state observer for the system

$$S_1: \quad \begin{aligned} x_1' &= x_2 \\ x_2' &= -x_1 + 2x_2 + u \\ y &= x_1 \end{aligned}$$

of Example 18.3 ($c_1 = 1$ and $c_2 = 0$). The observer will be

$$S_2: \quad \begin{aligned} w_1' &= e_{11}w_1 + e_{12}w_2 + f_1x_1 + g_1u \\ w_2' &= e_{21}w_1 + e_{22}w_2 + f_2x_2 + g_2u; \end{aligned}$$

$A = \begin{pmatrix} 0 & 1 \\ -1 & 2 \end{pmatrix}$, $B = \begin{pmatrix} 0 \\ 1 \end{pmatrix}$; $C = (1 \quad 0)$, $E = \begin{pmatrix} e_{11} & e_{12} \\ e_{21} & e_{22} \end{pmatrix}$,

$F = \begin{pmatrix} f_1 \\ f_2 \end{pmatrix}$, and $G = \begin{pmatrix} g_1 \\ g_2 \end{pmatrix}$. Hence $G = B = \begin{pmatrix} 0 \\ 1 \end{pmatrix}$,

$E = A - FC = \begin{pmatrix} -f_1 & 1 \\ -f_2-1 & 2 \end{pmatrix}$, and the characteristic polynomial of E is $\lambda^2 + (f_1-2)\lambda + 1+f_2 - 2f_1$. Taking $f_1 = 2$ and $f_2 = 3$, we obtain

$$E = \begin{pmatrix} -2 & 1 \\ -4 & 2 \end{pmatrix} \quad \text{and} \quad E^2 = 0.$$

The equation for the observer is

$$S_2: \quad \begin{aligned} w_1' &= -2w_1 + w_2 + 2x_1 \\ w_2' &= -4w_1 + 2w_2 + 3x_1 + u, \end{aligned}$$

and, for $n \geq 2$, $w_1(n) = x_1(n)$ and $w_2(n) = x_2(n)$; within time 2 the observer is exact, and this does not depend upon the estimates $w_1(0)$ and $w_2(0)$ of the initial state. With linear feedback $u = h_1w_1 + h_2w_2$, the output of the observer can be used to stabilize the system, since the system is also completely controllable.

18.7. <u>Exercise</u>. Using the observer in the above example and linear feedback $u = h_1w_1 + h_2w_2$, determine h_1 and h_2 so that the linear feedback control brings every initial state to the origin in finite time. Draw a block diagram of your resulting feedback control system.

We know from Proposition 18.5 that when S is observable we can make all of the eigenvalues of E zero. Then $E^m = 0$ and $w(n) = x(n)$ for all $n \geq m$. Within time m the observer is in theory exact. Now this state observer is always of dimension m, and the example we looked at earlier (Example 18.3) did suggest we can estimate the state with observers that are simpler to construct; i.e., by observers of lower dimension. We now want to look at this question in general for equation (18.6).

Let C be an $s \times m$ matrix and K an $s_0 \times m$ matrix. The dimension of the output y of S is s and the dimension of the output w of the observer S_0 is s_0. Let us suppose that S_0 observes Kx. Then $\begin{pmatrix} y(n) \\ w(n) \end{pmatrix} - \overline{K}x(n) \to 0$ as $n \to \infty$ where $\overline{K} = \begin{pmatrix} C \\ K \end{pmatrix}$. Now $\overline{K}$ is an $(s+s_0) \times m$ matrix and defines a linear transformation of R^m into R^{s+s_0}. This transformation is one-to-one on the image of $\overline{K}$ if and only if rank $\overline{K}$ = dim(image $\overline{K}$) = m; i.e., given $\overline{y} \in$ image K, there is a unique $x \in R^m$ such that $\overline{y} = \overline{K}x$. Define $x = \overline{K}^{-1}\overline{y}$ so that

$$\overline{K}^{-1}\begin{pmatrix} y(n) \\ w(n) \end{pmatrix} - x(n) \to 0$$

as $n \to \infty$, and we have an estimate of the state of S. In this case we will also call S_0 a <u>state observer</u>.

We may always assume that rank $C = s$ -- select the maximal number of linearly independent outputs y_i corresponding to linearly independent rows of C and discard the remaining outputs (columns of C). Assume that S_0 is a state observer so that rank $\overline{K} = m$. Then by properly numbering of the components of w we may assume that the rows of C plus the first $(m-s)$ rows of K are linearly independent. Let K_1 be the matrix consisting of the first $(m-s)$ rows of K; $K_1 = (I_{m-s}, 0)K$, where I_j is the $j \times j$ identity matrix. Then $\overline{K} = \begin{pmatrix} C \\ K_1 \end{pmatrix}$ is $m \times m$ and nonsingular, and $\overline{K}^{-1}\begin{pmatrix} y(n) \\ w(n) \end{pmatrix} - x(n) \to 0$ as $n \to \infty$; $\overline{w}(n) = \overline{K}^{-1}\begin{pmatrix} y(n) \\ w(n) \end{pmatrix}$ estimates the state. Since we, in effect, use only $m-s$ of the outputs of the observer S_0 we should be able to construct an $(m-s)$-dimensional state observer.

Again, looking at a simple example of the construction of a lower dimensional state observer shows us fairly well what can be expected in general. Consider

$$S: \quad \begin{aligned} x_1' &= -x_1 + x_2 \\ x_2' &= x_1 + \alpha x_3 + u_1 \\ x_3' &= 3x_1 + \beta x_3 + u_2 \\ y_1 &= x_1 + x_2 \\ y_2 &= x_1 - x_2 \\ y_3 &= x_1; \end{aligned}$$

$$A = \begin{pmatrix} -1 & 1 & 0 \\ 1 & 0 & \alpha \\ 3 & 0 & \beta \end{pmatrix}, \quad B = \begin{pmatrix} 0 & 0 \\ 1 & 0 \\ 0 & 1 \end{pmatrix}, \quad C = \begin{pmatrix} 1 & 1 & 0 \\ 1 & -1 & 0 \\ 1 & 0 & 0 \end{pmatrix}.$$

The output matrix is of rank 2. The outputs are not linearly independent, and by omitting y_2 and renumbering an equivalent output is

$$y_1 = x_1$$
$$y_2 = x_1 + x_2,$$

which corresponds to

$$C = \begin{bmatrix} 1 & 0 & 0 \\ 1 & 1 & 0 \end{bmatrix}.$$

Since

$$CA = \begin{bmatrix} -1 & 0 & 0 \\ 0 & 1 & \alpha \end{bmatrix},$$

the rank of $\begin{bmatrix} C \\ CA \end{bmatrix}$ is 3 if $\alpha \neq 0$, and hence this system S is observable if and only if $\alpha \neq 0$. Let us now try to construct an observer for $y_3 = r_1x_1 + r_2x_2 + x_3$ -- r_1 and r_2 to be selected later. Here $K = (r_1, r_2, 1)$ and

$\text{rank}\binom{C}{K} = 3$ for all r_1 and r_2. If we knew y_3, we would know the state x -- y_1, y_2 and y_3 are linearly independent. We let

$$w' = ew + f_1y_1 + f_2y_2 + g_1u_1 + g_2u_2,$$

and to be an observer for y_3 we want

$$w' - y_3' = e(w-y_3), \quad |e| < 1.$$

Since

$$w'-y_3' = ew + (f_1+f_2-r_1-r_2-3)x_1$$
$$+ (f_2-r_1)x_2 - (\alpha r_2+\beta)x_3 + (g_1-r_2)u_1$$
$$+ (g_2-1)u_2,$$

the equations to be satisfied are

$$\begin{aligned} f_1+f_2-r_1-r_2-3 &= -er_1 \\ f_2-r_1 &= -er_2 \\ \alpha r_2+\beta &= e \\ g_1-r_2 &= 0 \\ g_2-1 &= 0. \end{aligned}$$

If the system is observable ($\alpha \neq 0$), we can arbitrarily assign e; for instance, with $r_2 = -\frac{\beta}{\alpha}$ we have $e = 0$ and can take $g_1 = r_2$, $g_2 = 1$, $r_1 = 0$, $f_2 = 0$, and $f_1 = 3+r_2$. Then the observer is

$$w' = (3-\frac{\beta}{\alpha})y_1 - \frac{\beta}{\alpha} u_1 + u_2,$$

and

$$w(n) = -\frac{\beta}{\alpha} x_2(n) + x_3(n), \quad n \geq 1.$$

If the system is not observable ($\alpha = 0$), $e = \beta$ and, if $|\beta| < 1$,

$$w' = \beta w + 3y_1 + u_2$$

observes x_3; $w(n) = x_3(n) + \beta^n(w(0)-x_3(0))$ is an asymptotic estimate of $x_3(n)$. This example suggests, and we will see in a moment that this is true, that observability of S is a sufficient condition for being able to construct a state observer S_0 of dimension $m-s$ and shows that the condition is not necessary.

18.8. Exercise. For the system and the observer described above with $\alpha = 0$ (the system is not observable) and $|\beta| < 1$, show that the system can be stabilized using feedback control $u_1 = h_1x_1 + h_2(x_1+x_2) + h_3w$. In fact, for the resulting system

$$x_1' = -x_1 + x_2$$
$$x_2' = (1+h_1+h_2)x_1 + h_2x_2 + h_3w$$
$$x_3' = 3x_1 + \beta x_3$$
$$w' = \beta w + 3x_1 ,$$

show that β is always an eigenvalue and the other eigenvalues can be arbitrarily assigned. Determine h_1, h_2 and h_3 so that the other eigenvalues are zero.

We turn now to the general problem of constructing a lower order observer for

$$S: \quad \begin{aligned} x' &= Ax + Bu \\ y &= Cx. \end{aligned} \tag{18.9}$$

It is convenient to note first that we can always assume that S is of the form

$$S: \quad \begin{aligned} zx^1 &= A_{11}x^1 + A_{12}x^2 + B_1u \\ zx^2 &= A_{21}x^1 + A_{22}x^3 + B_2u \\ y &= x^1, \end{aligned} \tag{18.10}$$

where x^1 is s-dimensional, x^2 is (m-s)-dimensional,

$$A = \begin{pmatrix} A_{11} & A_{12} \\ A_{21} & A_{22} \end{pmatrix}, \quad B = \begin{pmatrix} B_1u \\ B_2u \end{pmatrix}, \quad C = (I_s, \ 0), \quad x = \begin{pmatrix} x^1 \\ x^2 \end{pmatrix}$$

and I_s is the $s \times s$ identity matrix. As we pointed out earlier we can always assume that C is $s \times m$ and rank $C = s$ -- the outputs are linearly independent. Also by numbering the outputs properly, we may assume that $C = (C_1, C_2)$ where C_1 is $s \times s$ and nonsingular. Make the change of coordinates $x = Q\overline{x}$ where

$$Q^{-1} = \begin{pmatrix} C_1 & C_2 \\ 0 & I_{m-s} \end{pmatrix} \quad \text{and} \quad Q = \begin{pmatrix} C_1^{-1} & -C_1^{-1}C_2 \\ 0 & I_{m-s} \end{pmatrix} .$$

Then

$$y = CQ\bar{x} = (I_s, 0)\bar{x} = \bar{x}^1,$$

the first s components of $\bar{x}$. In these new coordinates S is of the form (18.10) and we drop the bar over x.

Since the output of S is x^1, we would know the state if we knew $x^2 + Rx^1$ where R is any $(m-s) \times s$ matrix. We want therefore to construct an observer

$$S_0: \quad w' = Ew + Fy + Gu$$

that observes $x^2 + Rx^1$. If we can do this, then S_0 is an $(m-s)$-dimensional state observer. We want therefore to select E, F, G and R so that

$$z(w-x^2-Rx^1) = E(w-x^2-Rx^1)$$

and E is stable; i.e., we want

$$Ew + Fx^1 + Gu - [A_{21}x^1 + A_{22}x^2 + B_2u$$
$$+ R(A_{11}x^1 + A_{12}x^2 + B_1u)] = E(w-x^2-Rx^1).$$

Hence selecting

$$E = A_{22} + RA_{12}$$
$$F = A_{21} + RA_{11} - ER$$
$$G = B_2 + RB_1,$$

we see that S_0 observes $x^2 + Rx^1$ if there is an R that makes E stable. We know from Corollary 18.2 that a sufficient condition for this is that the pair $\{A_{22},A_{12}\}$ be observable, and, if so, we can arbitrarily assign the eigenvalues of E.

18.8. <u>Proposition</u>. $\{A,C\}$ observable implies $\{A_{22},A_{12}\}$

observable.

Proof: Note that

$$CA = (I_s, 0)\begin{pmatrix} A_{11} & A_{12} \\ A_{21} & A_{22} \end{pmatrix} = (A_{11}, A_{12})$$

$$CA^2 = (A_{11}, A_{12})\begin{pmatrix} A_{11} & A_{12} \\ A_{21} & A_{22} \end{pmatrix}$$

$$= (A_{11}^2 + A_{12}A_{21}, \quad A_{11}A_{12} + A_{12}A_{22}).$$

Letting

$$CA^n = (C_1^{(n)}, \quad C_2^{(n)}),$$

it is not difficult to see that

$$C_2^{(n)} = \sum_{j=0}^{n-1} K_{n,j}A_{12}A_{22}^j,$$

since $C_2^{(n+1)} = C_1^{(n)}A_{12} + C_2^{(n)}A_{22}$ and therefore $K_{n+1,0} = C_1^{(n)}$ and $K_{n+1,j+1} = K_{n,j}$; for $j = 0,\ldots,n-1$. Now $\{A_{22},A_{12}\}$ not observable implies there is an $x^2 \neq 0$ such that $A_{12}A_{22}^n x^2 = 0$ for all $n \geq 0$. But then $CA^n\begin{pmatrix} 0 \\ x^2 \end{pmatrix} = \begin{pmatrix} 0 \\ C_2^{(n)}x^2 \end{pmatrix}$ $= 0$ for all $n \geq 0$, and $\{A,C\}$ is not observable. □

This proposition plus what was shown above proves the following result.

18.9. Theorem. If an m-dimensional linear system S with s linearly independent outputs (rank C = s) is observable, then an (m-s)-dimensional state observer S with (E having) preassigned eigenvalues can be constructed.

18.10. Exercise. Construct whenever possible for the system S in Example 18.3 a one-dimensional observer with zero eigenvalues.

Let us consider the problem of stabilizing the system

$$S: \quad \begin{aligned} x' &= Ax + Bu \\ y &= Cx \end{aligned}$$

using linear feedback under the assumption that we can construct an $(m-s)$-dimensional state observer

$$S_0: \quad w' = Ew + Fy + Gu;$$

C is $m \times s$ and rank $C = s$. Then S_0 observes Kx and $\overline{K} = \binom{C}{K}$ is $m \times m$ and nonsingular; i.e., E is stable and $\begin{pmatrix} y(n) \\ w(n) \end{pmatrix} - \overline{K}x(n) \to 0$ as $n \to \infty$. If S is observable, we can always construct S_0. Let us assume also that

$$x' = Ax + Bu$$

can be stabilized by linear feedback $u = \Gamma x$; i.e., there is Γ such that $A + B\Gamma$ is stable. A sufficient condition for this is that S be controllable. Thus our assumptions are satisfied if S is both controllable and observable, and in this case we can preassign the eigenvalues of $A + B\Gamma$ and E. Under our assumptions we should certainly expect to be able to stabilize S using linear feedback $u = H_1y + H_2w$ (see Figure 18.7). Engineers call this "dynamic" feedback. A difference equation is used to generate the control.

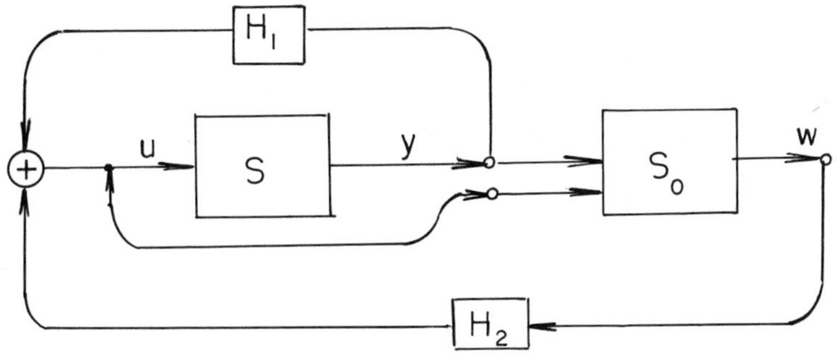

Figure 18.7

Since

$$H_1y(n) + H_2w(n) - (H_1C + H_2K)x(n) \to 0,$$

we see by selecting $H_1C + H_2K = \Gamma$, the control is asymptotic to the stabilizing control Γx;

$$H_1C + H_2K = (H_1,H_2)\overline{K},$$

and we select

$$(H_1,H_2) = \Gamma\overline{K}^{-1}.$$

The equations of this feedback system are then

$$x' = (A+BH_1C)x + BH_2w$$
$$w' = (FC+GH_1C)x + (E+GH_2)w.$$

Writing these equations in terms of the more natural co-ordinates $\varepsilon = w - Kx$ and x, we obtain

$$x' = (A+BH_1C+BH_2)x + BH_2\varepsilon$$
$$\varepsilon' = E\varepsilon$$

or

$$x' = (A+B\Gamma)x + BH_2\varepsilon$$
$$\varepsilon' = E\varepsilon.$$

The matrix of this system is

$$\begin{pmatrix} A+B\Gamma & BH_2 \\ 0 & E \end{pmatrix},$$

its eigenvalues are those of $A+B\Gamma$ and E; its characteristic polynomial is $\det(\lambda I-A-B\Gamma)\cdot\det(\lambda I-E)$. This composite system is stable if and only if $A+B\Gamma$ and E are stable.

We have therefore the following result (see Theorems 18.9 and 16.4).

18.11. Theorem. Given an m-dimensional linear system S with s linearly independent outputs that is both controllable and observable it is always possible to construct an $(m-s)$-dimensional dynamic feedback system with the $2m-s$ eigenvalues of the composite system arbitrarily assigned.

Under the hypotheses in the above all of the eigenvalues can be made to vanish, and hence there is a dynamic feedback control that brings each initial state to the origin in time $2m-s$; in time $m-s$ the state is identified and in time m the control brings the system to the origin.

REFERENCES

Arbib, M. and Padulo, L., [1], System Theory: A Unified State-Space Approach to Continuous and Discrete Systems, Hemisphere Pub., New York, 1974.

Bellman, R., Introduction to Matrix Analysis, McGraw-Hill, New York, 1960.

Ben-Israel, A. and Greville, T., Generalized Inverses: Theory and Applications, Wiley-Interscience, New York, 1974.

Fuller, A. T., [1], The early development of control theory, II, J. Dynamic Systems, Measurement, and Control, Trans. ASME, 98, 2, 1976, pp. 224-234.

Gantmacher, F. R., [1], The Theory of Matrices I & II, Chelsea Publ., New York, 1960.

Jury, E. I., [1], Inners and Stability of Dynamical Systems, Robert E. Krieger Publ., Malabar, Florida, 1982.

Langenhop, C. E., [1], On the stabilization of linear systems, Proc. Amer. Math. Soc., 15, 5, 1964, pp. 735-42.

LaSalle, J. P., [1], Stability theory for difference equations. Studies in Ordinary Differential Equations, MAA Studies in Math., Amer. Math. Assoc., 1977, pp. 1-31.

LaSalle, J. P., [2], The Stability of Dynamical Systems, SIAM CMBS 25, 1976.

Maxwell, J. C., [1], On governors, Proc. Roy. Soc. London, 16, 1868, pp. 27-283.

Nering, E., [1], Linear Algebra and Matrix Theory, J. Wiley, New York, 1963.

Perron, O., [1], Über Stabilität und asymptotische Verhalten der Lösungen eines Systems endlicher Differenzengleichungen, J. Reine Angew. Math., 161, 1929, pp. 41-61.

Wonham, W. M., [1], On pole assignment in multi-input controllable linear systems, IEEE Trans., AC-12, 6, 1967, pp. 660-665.

INDEX

Applied Mathematical Sciences

cont. from page ii

39. Piccini/Stampacchia/Vidossich: **Ordinary Differential Equations in R^n.**
40. Naylor/Sell: **Linear Operator Theory in Engineering and Science.**
41. Sparrow: **The Lorenz Equations: Bifurcations, Chaos, and Strange Attractors.**
42. Guckenheimer/Holmes: **Nonlinear Oscillations, Dynamical Systems and Bifurcations of Vector Fields.**
43. Ockendon/Tayler: **Inviscid Fluid Flows.**
44. Pazy: **Semigroups of Linear Operators and Applications to Partial Differential Equations.**
45. Glashoff/Gustafson: **Linear Optimization and Approximation: An Introduction to the Theoretical Analysis and Numerical Treatment of Semi-Infinite Programs.**
46. Wilcox: **Scattering Theory for Diffraction Gratings.**
47. Hale et al.: **An Introduction to Infinite Dimensional Dynamical Systems—Geometric Theory.**
48. Murray: **Asymptotic Analysis.**
49. Ladyzhenskaya: **The Boundary-Value Problems of Mathematical Physics.**
50. Wilcox: **Sound Propagation in Stratified Fluids.**
51. Golubitsky/Schaeffer: **Bifurcation and Groups in Bifurcation Theory, Vol. I.**
52. Chipot: **Variational Inequalities and Flow in Porous Media.**
53. Majda: **Compressible Fluid Flow and Systems of Conservation Laws in Several Space Variables.**
54. Wasow: **Linear Turning Point Theory.**
55. Yosida: **Operational Calculus: A Theory of Hyperfunctions.**
56. Chang/Howes: **Nonlinear Singular Perturbation Phenomena: Theory and Applications.**
57. Reinhardt: **Analysis of Approximation Methods for Differential and Integral Equations.**
58. Dwoyer/Hussaini/Voigt (eds.): **Theoretical Approaches to Turbulence.**
59. Sanders/Verhulst: **Averaging Methods in Nonlinear Dynamical Systems.**
60. Ghil/Childress: **Topics in Geophysical Fluid Dynamics: Atmospheric Dynamics, Dynamo Theory and Climate Dynamics.**
61. Sattinger/Weaver: **Lie Groups and Algebras with Applications to Physics, Geometry, and Mechanics.**
62. LaSalle: **The Stability and Control of Discrete Processes.**

ISBN 0-387-96411-8
ISBN 3-540-96411-8